DESIGN OF WASTEWATER AND STORMWATER PUMPING STATIONS

WEF Manual of Practice FD-4

Third Edition

Edited by Robert Otey, PE, PACP, and
Mazen Kawasmi, PE, CFM, GISP, PACP, IAM

2022

Water Environment Federation
601 Wythe Street
Alexandria, VA 22314–1994 USA
http://www.wef.org

ISBN 978-1-57278-436-9

About WEF

The Water Environment Federation (WEF) is a not-for-profit technical and educational organization of 30,000 individual members and 75 affiliated Member Associations representing water quality professionals around the world. Since 1928, WEF and its members have protected public health and the environment. As a global water sector leader, our mission is to connect water professionals; enrich the expertise of water professionals; increase the awareness of the impact and value of water; and provide a platform for water sector innovation. To learn more, visit www.wef.org.

Third edition edited by Robert Otey, PE, PACP, and Mazen Kawasmi, PE, CFM, GISP, PACP, IAM

With contributions by:

Jared Barber, PE, PACP, MACP, LACP
Lewis Bernard, PE
Jeremy Dixon, PE, CFM
Jonathan Faughtenberry, PE
Erin Flanagan, PE, BCEE
Scott Haberstroh, PE
Jeff Hensley, PE
William (Clay) Herndon, PE
Thomas Hill, PE
Scott Hubley, PE, CFM
Alan Hutson, PE, ENV SP
Bryan Jann, PE
Charles Kucherka, PE, CCM, NACE CIP-3
Aubrey Monroe
Rebecca Peters, PE
Tony Pevec, PE, BCEE, DBIA
Nick Prisco, PE
Rebecca Sandoval, PE
Tarlton W. Smith II, PE, DBIA, ENV SP
Brogan Tyler
Scott Vaughan

Review provided by:

Chein-Chi Chang, PhD, PE
Charles Falzone, PE, PSAP
James Gagnon
Michael Jankowski, PE
Michel Wanna

Text originally prepared by the **Design of Wastewater and Stormwater Pumping Stations Task Force** of the **Water Environment Federation**

Terry L. Krause, *Chair*

Diego Alaimo
Richard L. Anderton
Robert J. Andryszak
Peter J. Barthuly
Henry Benjes Jr.
Aivars Bergs
Douglas G. Brinkman
Gerald T. Carey
Steven W. Collins
Tracy O. Dubcak
Joseph Gemin
Frank M. Gunby Jr.
Paul J. Higgins
Barton G. Jones
Madhusudan C. Joshi
Louis W. Klinefelter
J. Robert Krebs
James D. Linsley
Chung-Lyu Liu
Michael L. Middleton
Darrell Morrison
Roy Parsons
Manu A. Patel
Calvin C. Patterson
Charles M. Pekkala
Thomas A. Ridgik
Paul R. Rom
Thomas M. Simpson
Adam H. Slutsky
Peter M. Smith
Wilson F. So
Marc S. Solomon
Srinivasan Sundaramoorthy
Thomas G. Thompson
Rufino A. Vigilia Jr.
Thomas M. Walski
Thomas L. Weber
Glenn R. Wentink
H. Leroy Whiteley Jr.

Under the Direction of the **Municipal Resource Recovery Design Subcommittee** of the **Technical Practice Committee**

Manuals of Practice of the Water Environment Federation

The WEF Technical Practice Committee (formerly the Committee on Sewage and Industrial Wastes Practice of the Federation of Sewage and Industrial Wastes Associations) was created by the Federation Board of Control on October 11, 1941. The primary function of the Committee is to originate and produce, through appropriate subcommittees, publications dealing with technical aspects of the broad interests of the Federation. These publications are intended to provide background information through a review of technical practices and detailed procedures that research and experience have shown to be functional and practical.

Contents

List of Figures

List of Tables

Preface

This manual of practice presents contemporary design considerations for wastewater and stormwater pumping stations. The manual, which updates the previous pumping station design manual published in 1993 by the Water Environment Federation, covers station capacity requirements and configuration, the selection and design of pumping equipment, piping system layout and hydraulic considerations, electrical and instrumentation systems design, and other station appurtenances.

The manual is intended for the practicing design professional. Accordingly, it focuses on current design practice for those individuals that have a general knowledge of the basic principles of pumping station design. Considerations discussed are intended to cover the design of both small and large pumping stations and highlight the differences between wastewater and stormwater stations.

The contents of this manual represent the collected background and experience of more than 20 professionals active in the design of wastewater and stormwater pumping stations. Material presented is intended to complement other Water Environment Federation manuals of practice and published design guidelines and textbooks.

The third edition of this Manual of Practice was edited under the direction of Robert Otey, PE, PACP, and Mazen Kawasmi, PE, CFM, GISP, PACP, IAM.

Authors' and reviewers' efforts were supported by the following organizations:

Chang Tech International, Inc.
Drummond Carpenter, PLLC
Freese and Nichols, Tampa, Florida; New Orleans, Louisiana; Charlotte, North Carolina; Raleigh, North Carolina; Tulsa, Oklahoma; Austin, Texas; Dallas, Texas; Fort Worth, Texas; Houston, Texas; Oklahoma City, Texas; Pearland, Texas; San Marcos, Texas
Macon Water Authority
Orange County Sanitation District

1

Introduction

1.0 OVERVIEW

Wastewater and stormwater collection and conveyance systems are extensive networks of interconnected piping. Because of the expansiveness and topography of these systems, wastewater or stormwater flows cannot always be conveyed by gravity. As a result, pumping stations are needed to lift the flow over high points in the systems or to treatment facilities. Because these stations play a vital role in the collection system, design engineers must produce plans for well-engineered pumping stations that operate reliably and efficiently.

Wastewater or stormwater pumping stations range from simple facilities of low capacity in remote areas to complex facilities capable of pumping billions of liters (gallons) of wastewater or stormwater each day. Often, especially at the influent end of a treatment facility, the pumping station is combined with other preliminary treatment systems. As a result, the design of these facilities is a complex undertaking that requires an understanding of all of the design engineering disciplines. The design engineer should use all the technical disciplines early in the design process to ensure a well-coordinated and functional pumping station.

Originally published in 1993 by the Water Environment Federation, this manual has been prepared to assist the practicing engineer in the design of wastewater and stormwater pumping stations; it is for the experienced designer rather than the novice. While there are many similarities between the design of wastewater and stormwater pumping stations, there also are differences. For example, while there is similarity between wastewater and stormwater pumps, pipes, valves, and controls, there are differences in station hydraulics and operational considerations such as the intermittent operation of a stormwater pumping station. This manual has been revised to include updated processes and technological advances in pumping stations. For a detailed description of those changes, refer to the chapter descriptions located within Section 2.0.

2.0 GUIDE TO THE MANUAL

The content and organization of the manual is intended to assist in the planning and design of both wastewater and stormwater pumping stations. The manual is divided into eight chapters and should be used in conjunction with other Water Environment Federation manuals that deal with the design of collection and treatment facilities.

Chapter 2, "Station Capacity," covers the methods of determining the need for a pumping station and the station's capacity. Determining the peak hourly wastewater flowrate and developing an inflow hydrograph for determining stormwater flowrate also are discussed. Updates to this chapter include computational fluid dynamics descriptions, inclusion of Ten States Standards, inflow/infiltration modeling, volume sizing determinations, effects of inflow hydrographs on the system, analysis of more scenarios, and evaluating parameters such as peak stage, peak discharge, drain time, and pump cycling.

Chapter 3, "Station Configuration and Design," presents information on facility location, types, aesthetic and safety considerations, and architectural and structural design. Wet well configuration and storage considerations also are addressed. Updates to this chapter include identifying the distinction between wastewater and stormwater stations and describing larger stormwater pumping stations and the unique challenges faced on the coast and/or during major storm events.

Chapter 4, "Pumping Systems," reviews pumping system definitions and describes the types of pumps and drives that are typically used. The chapter also presents pumping system design concerns such as pump curve preparation and pump selection criteria. Updates to this chapter include a more in-depth discussion of "allowable operating range" and "preferred operating range" designations.

Chapter 5, "Piping Systems," reviews fluid flow fundamentals, water hammer and surge considerations, and the design of station piping systems, including piping arrangements, piping materials, valves, and flow-metering considerations. Updates to this chapter include a more in-depth discussion of piping requirements for stormwater pumping stations; expansions on flow meters, fusible pipe, and *C* factors; and hydraulics such as evaluating minor losses.

Chapter 6, "Electrical Design," describes the layout of a pumping station's electrical system, including reliability and redundancy, voltage considerations, substations, transformers, switchgears, conduit and wire, lightning protection, standby power considerations, and adjustable-speed drives. Updates to this chapter include revised references to National Fire Protection Association and IEEE standards such as National Electrical Code, National Electrical Safety Code, IEEE Color Books, evaluating arc flash hazards, and discussion of the several types of variable frequency drive technologies.

Chapter 7, "Instrumentation and Control Systems," presents information about process and instrumentation diagrams, flow gauges, flow metering elements, system alarms, controls, and telemetry systems. Updates to this chapter include revised instrumentation options such as radar for level measurement, fiber optics for supervisory control and data acquisition systems, and evaluating options for communication between different electrical, control, and instrumentation equipment via Ethernet.

Chapter 8, "Appurtenances," discusses design considerations for typical appurtenances found in pumping stations, including screening, odor control and treatment, noise control, and building services such as heating, ventilation and cooling, lighting, water supply, drainage, utilities, hoists, and safety features. Updates to this chapter include incorporating the latest technology in screening equipment, addressing non-dispersibles, and evaluating the pros and cons of odor control options.

2

Station Capacity

1.0 SANITARY WASTEWATER FLOW

Sanitary wastewater primarily consists of used portions of public or private potable water supplies that are discharged to sewers through the plumbing systems of residential buildings. It may also include portions of a public or private water supply discharged through the plumbing systems of industrial and commercial establishments. Sanitary wastewater also contains an infiltration and inflow (I/I) component. The ratio between residential, industrial, and commercial and I/I components will vary from one sanitary waste collection system to another. Before final discharge, these flows typically are collected and carried to a treatment facility where various pollutants are reduced or removed.

To provide economical and effective treatment for sanitary wastewaters, their volume should be kept to a minimum and not combined and diluted with larger volume stormwater flows, infiltration, or inflow. Inflow, such as storm runoff from roof, yard, and foundation drains, typically is collected separately from sanitary wastewater and conveyed to a disposal point by a separate sewer or drainage system designed for this purpose.

1.1 Engineering Report

Before the design and construction of a sanitary wastewater collection and pumping station, an engineering report detailing the basis of design should be prepared. This report should analyze associated planning studies, drainage areas, population growth projections, land development, quantity of available water, projected quantity of wastewater flows—including all significant wastewater contributors—I/I, and the quality of influent water. The engineering report often includes the design schedule for the project and estimated costs of alternatives.

Based on the results of these preliminary studies, the engineering report should be expanded, if applicable, to include the pumping station basis of design. The basis of design should include the range of expected flows for which the facility will be designed, the number and capacity of pumps, the design life, and the design criteria for each engineering discipline (that is, civil, structural, architectural, mechanical, electrical, and instrumentation and control) involved. The basis of design should include an estimate of the construction cost, a preliminary assessment of the type of equipment, need for screening or other pretreatment, reliability standards (standby pumps and backup power supplies), need for station amenities (such as restrooms and offices), general facility layout, equipment access, control system, monitoring needs, overall pumping station aesthetics, and odor control considerations (Jones & Sanks, 2008).

This chapter addresses development of the facility design life and determination of facility capacity. Subsequent chapters address the facility basis of design.

1.2 Design Life

There is a maximum hydraulic or functional capacity beyond which a wastewater system will not operate satisfactorily. The time span, a period ranging from 10 to 30 years from the date of the original design to an estimated future date when the capacity of the system will be reached, is referred to as the "design life." The system's design life should be determined before the start of any engineering design. An analysis of the design life should consider the following:

- Anticipated obsolescence and wear and tear on the mechanical equipment,
- The possibility of needing to expand the system in the future,
- The ease or difficulty of implementing such an expansion, and
- The efficiency of the system's performance during its initial years when it may not yet be operating to full capacity.

In choosing a design life, undue financial burden on the system's present users should be avoided. A longer design life will require that larger facilities be constructed, resulting in higher construction and operations and maintenance (O&M) costs. Selecting too short a design life could result in additional facility expansion costs sooner than desired. Careful selection of the design life is important to ensure that the system's cost be shared equitably by existing and future users.

In summary, selecting a reasonable design life should reflect the ability of the design to accommodate the flexibility to meet future conditions through planned modification options (impeller and motor changes) or expansion of standard elements (pumps).

1.3 Population Growth Projection

The quantity of sanitary wastewater generated by a community depends largely on the population and the per capita wastewater contribution. Therefore, an accurate population projection is essential to implementing a well-planned collection and treatment system. A community's population growth and distribution may be affected by local economics, employment opportunities, commuting distances for workers, socioeconomic factors, zoning, availability of community facilities and services, and other community amenities. Factors not evident at the time of the study may emerge to cause a major shift in population growth, for example, the discovery of new natural resources in the vicinity, availability of new or expanded water supplies, or an industry's decision to move to or from the community.

Historical population information can be obtained from various sources. The U.S. government decennially has conducted a census since 1790. The census provides valuable demographic information on population, such as socioeconomic characteristics, housing facts, and population per household.

Additional data for the interim period may be obtained from local surveys for schools, building permit records, tax records, chamber of commerce studies, local public utility records, and local and regional planning commissions. Careful analysis of census data may reveal past population trends.

There are various methods of projecting population trends, depending on the amount and types of historic data available. These are determined by the following factors:

- Arithmetic (changes population by a fixed constant quantity);
- Geometric (changes population in proportion to population size, by a fixed ratio);
- Cohort survival (determines population by age groups and sex);
- Graphical (comparison with the growth of a similar but larger community);
- Apportionment (establishes population based on projections for larger dimensions, such as county or state); and
- A combination of the above.

Because some of these methods may be complicated, especially when used in combination with one another, the desired degree of accuracy should be considered when projection techniques are selected. The engineer should check with the local planning agency first, because in many cases population projections may already have been developed in sufficient detail. Population projections should be reviewed with the local planning agency and the project owner before proceeding with the study.

Details of population projection methods are discussed in many textbooks and references (McJunkin, 1964; McLean, 1952; United Nations, n.d.). Most graphical or mathematical projections are based on extending past growth trends. These methods are more applicable to communities with historically stable growth. For young and fast-growing communities, or communities with high growth potential, a meaningful projection requires good engineering judgment based on past information coupled with consideration of all related environmental determinants. Similarly, consideration should be given to the possibility of reduced growth rates in older or more mature communities at or near growth saturation.

Population projections can also be obtained from previous studies prepared for the community. Previous U.S. Environmental Protection Agency (U.S. EPA)–funded construction grant studies contain population projections for the community or for larger basin areas. These projections may require further breakdown for meaningful use on a single project covering a smaller area.

1.4 Land Development

Local and regional land use planning, development, urban renewal projects, and zoning regulations will influence future population growth and distribution, which, in turn, will affect the quantity of wastewater generated by a community. In many urban areas, burgeoning construction of multi-dwelling apartments, condominiums, and town homes had led to increased population densities and wastewater quantities.

Many housing units are equipped with appliances (such as a dishwasher, garbage disposal unit, or clothes washer) that increase water consumption and the amount of sanitary wastewater generated. However, rising costs of water in some communities have tended to suppress water use, hence reducing wastewater flow.

1.5 Quantity of Available Water

The quantity of wastewater flow generated is closely related to the quantity of available water and its consumption rate. In addition, the public's general tendency to use water will vary depending on weather conditions, metered or unmetered water supplies, water rates, local conservation efforts, availability of public sewers, and the economic condition of the community.

Because wastewater consists primarily of used water, which is often metered, per capita water consumption can be established. From this information, it is possible to estimate the per capita quantity of wastewater reaching the sewer. Records of water usage (daily, monthly, and seasonally) typically are available from the water utility that provides water service to the study area. National and regional water usage rates may also be available. Annual fluctuations in water demands and, subsequently, wastewater flows also play an important role in the design of sewerage systems, hence in the required capacity of pumping stations.

1.6 Quantity of Wastewater Flow

The quantity of wastewater flow varies continuously in a wastewater collection system. There will be a wide fluctuation of flow throughout the day in any system in which sanitary wastewater is a significant contributor to the total flow. If wastewater facilities are not designed to accommodate these fluctuations, they can and will cause operational problems in collection and pumping systems and in treatment facilities.

For design purposes, it is necessary to estimate future wastewater flows based on existing conditions. State and municipal regulatory agencies often provide average per capita domestic wastewater design flows or curves of flow versus population to evaluate required pumping station capacity. If no such guidance is available and the system has no flow records, or if a complex sewer network makes actual measurement of flows impractical, design flows may be determined by either estimating and totaling the flows from the various contributors or by calculating total flow as a percentage of the total available water supply. In either case, careful study and judgment are needed to project wastewater flow because average flows have been shown to vary between 190 and 1000 L/cap·d (50 and 265 gpd/cap) (American Society of Civil Engineers [ASCE], Environmental and Water Resources Institute [EWRI], & Water Environment Federation [WEF], 2007).

Engineers typically are interested in the average daily flow, maximum and minimum daily flows, and peak hourly flow in a sanitary sewer collection system for a variety of different design purposes. For pumping stations, the capacity of required pumping equipment and the pumping station wet well volume will depend on either the peak hourly or maximum daily flow, or a combination of the two (i.e., the peak hour of the maximum day), whichever is greater. Peak hourly flow will determine the maximum required pumping capacity of the station. However, minimum daily flows also are important in pumping station design, not only to assess maximum residence time in wet wells, but also to determine required turndown capacity of the pumping equipment. Turndown capacity is the ratio of maximum to minimum pumping capacity of the pumping station (e.g., a required turndown capacity of 2:1 would indicate that the minimum pumping capacity at which the pumping station will have to operate effectively is 50% that of the maximum capacity).

1.7 Additional Contributors

In addition to domestic wastewater flows, contributions may come from commercial, industrial, institutional, and I/I sources.

1.7.1 Commercial

Wastewater flows from commercial districts often are estimated in m^3/ha·d (gpd/ac), based on information from existing similar establishments, water consumption, and comparative data from other cities. Flow allowances for sewer design can vary from 42 000 to more than 1 500 000 L/ha·d (4500 to 160 000 gpd/ac) in different cities and depend on the type of commercial activity (ASCE, EWRI, & WEF, 2007). Generous allowances for many commercial establishments may be justified because of the possibilities of future expansions and improvements.

1.7.2 Industrial

Two basic types of wastewater typically are generated by industries in addition to the sanitary flow originating from employees: process wastewater and spent cooling water. For existing industries, the quantity of process wastewater may be obtained through a field survey and flow monitoring or by analysis of water consumption records. For future industrial facilities, wastewater flow may be projected by studying the processes to be included in the new facility, the water consumption rate for each process, time of day and duration of process operation, types and numbers of industrial buildings, estimated number of employees, and comparative data from similar industries.

Spent cooling water to be handled by sanitary systems typically comes from water-cooled air-conditioning chiller units. The average water demand for once-through, water-cooled air-conditioning equipment ranges from 0.10 to 0.14×10^{-3} L/kg·s (1.5 to 2 gpm/ton) of refrigeration. Water for multiple-pass cooling systems and for major industrial processes generally is treated and handled by separate systems. A small amount of water generally is wasted or "blown down" to the sewer to maintain system stability. Consideration must be given to this water because, while it may be small in volume, it may contain a high proportion of certain pollutants such as water-conditioning additives and total dissolved solids.

1.7.3 Institutional

The quantity of institutional wastewater may be established based on such items as the population served, the number of buildings, the number of units of equipment, hygiene controls, and individual habits. The design flow can be determined by field survey of the existing facilities or by analyzing comparative data from similar institutions.

1.7.4 Infiltration and Inflow

When calculating design flows for pumping stations, consideration must be given to I/I. Inflow is water entering a sewer system and service connections from roof drain leaders, cellar drains, cooling water discharges, manhole covers, cross connections, and surface runoff (WEF & ASCE, 2018; WEF et al., 2009). Infiltration is groundwater that enters a sewer through defective and cracked pipes, open or loose pipe joint connections, and leaking manhole walls. The actual amount of I/I will depend on the material and integrity of the sewer system, the number and type of building connections, the elevation of the groundwater compared to that of the sewers, the integrity of manholes, and the degree of manhole submergence during a rainfall event. Infiltration and inflow to sanitary sewers most often is estimated in terms of the amount of water entering a sewer based on the diameter and length of pipe (mL/m·d)/m (gpd/in. dia/mi). Infiltration may range from 9260 to 926 000 (mL/m·d)/m (100 to more than 10 000 gpd/in. dia/mi) (ASCE, EWRI, & WEF, 2007).

In a properly designed, installed, and well-operated and maintained collection system, the I/I rate should not exceed 18 520 (mL/m·d)/m (200 gpd/in. dia/mi) for any section of the collection system (Commonwealth of Pennsylvania Department of Environmental Protection, 1997). Newer collection systems, constructed after 1970, should achieve I/I rates between 13 900 and 18 520 (mL/m·d)/m (150 and 200 gpd/in. dia/mi), whereas older systems with less efficient joint construction may experience leakage in excess of 27 780 (mL/m·d)/m (300 gpd/in. dia/mi).

Since the implementation of the Federal Water Pollution Control Act Amendments of 1972, which required grant recipients to demonstrate their sewer systems were not subject to excessive I/I, emphasis has continued to be placed on assessing and limiting I/I to sanitary sewer collection systems. Detailed pumping station design should rely on a more complete sewer system I/I evaluation that establishes accurate I/I flowrates through actual field measurement. While a cost-benefit analysis ultimately will determine whether the owner of a system should rehabilitate a sewer with excessive I/I, the designer of a pumping station needs to be able to assess the effect these additional flows will have on the design of the pumping facilities and, ultimately, treatment facilities.

Table 2.1 summarizes typical wastewater flows generated from both residential and nonresidential (commercial and industrial), institutional, and recreational (seasonal) establishments. Flows from these establishments can be added to determine the required pumping station capacity needed to serve an existing or new community.

1.8 Peak Hourly Wastewater Flowrate

The peak hourly wastewater flowrate is a function of the service area as well as the associated wastewater generation patterns and population being served and will determine the maximum required pumping station capacity. In the absence of actual flow date, the flowrate peaking factor, which is the ratio of peak hourly flowrate to average daily flowrate, can be determined

TABLE 2.1 Typical Wastewater Flows From Establishments Served by a Community Wastewater System (Pennsylvania DEP, 1983).

Type of Establishment	Wastewater Flow, gpd/person[a]
Residential	
Hotels and motels without private baths	40
Hotels and motels with private baths	50
Luxury residences and estates	125
Multiple-family dwellings and apartments	60
Rooming houses	50
Single-family houses	75
Commercial	
Airline catering per meal served	3
Airports (per passenger, not including food)	5
Airports (per employee)	10
Beauty shops (per operator chair)	200
Bus service areas, not including food	5
Country clubs, not including food	30
Drive-in theaters (per space per day, not including food)	10
Factories and plants (exclusive of industrial wastes)	35
Laundries, self-service (gallons/washer)	400
Mobile home parks, independent (per space)	250
Movie theaters (per auditorium seat, not including food)	5
Offices (per employee)	10
Restaurants (per patron, toilet, and kitchen wastes)	10
(Additional for bars and cocktail lounges)	2
Restaurants (single-service utensils/person, kitchen, and toilet wastes)	8.5
Restaurants (single-service utensils/patron, kitchen waste only)	3
Stores (per public toilet)	400
Warehouses	35
Work or construction camps (semipermanent) with flush toilets	50
Work or construction camps (semipermanent) without flush toilets	35

(*continued*)

TABLE 2.1 Typical Wastewater Flows From Establishments Served by a Community Wastewater System (Pennsylvania DEP, 1983). (*Continued*).

Type of Establishment	Wastewater Flow, gpd/ person[a]
Institutional	
Churches (per seat)	3
Churches (per meal served, kitchen waste)	3
Churches (per meal served, with paper service)	1.5
Hospitals (per bed space, with laundry)	300
Hospitals (per bed space, without laundry)	220
Institutional food service (per meal)	20
Institutions other than hospitals (per bed space)	125
Schools, boarding	100
Schools, day (without cafeterias, gyms, or showers)	15
Schools, day (with cafeterias, but no gym or showers)	20
Schools, day (with cafeterias, gyms, and showers)	25
Recreational and seasonal	
Camps, day (no meals served)	10
Camps, hunting and summer residential (night and day) with limited plumbing, including water-carried toilet wastes	50
Campgrounds, with individual sewer and water hookup (per space)	100
Campgrounds, with water hookup only and/or central comfort station that includes water-carried toilet wastes (per space)	50
Fairgrounds and parks, picnic with bathhouses, showers, and flush toilets	15
Fairgrounds and parks, picnic (toilet wastes only)	5
Swimming pools and bathhouses	10

Note. Actual flow data for any establishment indicating peak daily flows over a 1-year period may be accepted for use in sizing the on-lot disposal system.
[a]gpd × 3.785×10^{-3} = m^3/d

using historical flowrate records or estimated using the following equation (Fair & Geyer, 1954):

$$PF = \frac{18 + \sqrt{P}}{4 + \sqrt{P}} \tag{2.1}$$

Where

PF = peaking factor and

P = population in thousands.

The resulting peaking factor includes normal I/I for newer collection systems built with modern construction materials and methods. Higher peaking factors should be considered for collection systems that were built with older materials and construction methods or that are in areas susceptible to higher I/I rates, such as areas with high groundwater tables and areas with combined sewer systems. Pumping stations should be designed such that the peak hourly wastewater flow can be conveyed with the largest pump out of service.

2.0 STORMWATER FLOWS

The quantity of flow reaching a stormwater pumping station is determined by hydrologic analysis of regional rainfall events; the characteristics of the watershed; and the hydraulics of the collection, treatment, and discharge system (Chow, 1964; Chow et al., 1988; Linsley et al., 1958; Williams, 1950). In many instances, storm drainage systems are designed solely based on peak runoff values, thus neglecting flow variations with time and routing of the flow through the watershed, the collection and treatment systems, and any storage facilities along the flow route. This method is an adequate but conservative means of sizing gravity stormwater pipelines. The same simplified technique of using only peak flow values, however, cannot be applied to the design of stormwater pumping stations.

The design of stormwater pumping stations requires not only ascertaining the peak flow, but also defining the variation of flow with time to determine the design inflow to the pumping station. This curve of flow versus time is known as the *inflow hydrograph*. Methods for its development will be discussed later in this chapter. A distinction needs to be made that the stormwater hydrograph may not necessarily be the same as the pumping station inflow hydrograph. This is because the point of measurement for the stormwater event may be different from the location of the pumping station. As described previously for sanitary flows, an engineering design report also should be prepared for stormwater pumping stations. The report should address the system hydrology and the elements (such as existing and future watershed development, soil and cover type, and meteorological data) that will affect stormwater generation, methodology and development of the influent hydrograph, and selection of appropriate design flows.

2.1 Inflow Hydrograph

The inflow hydrograph must be defined to size the pumps and pumping station wet well properly and economically. For stormwater pumping stations, it is advisable to size pumps to accommodate the hourly peak inflow or provide variable pumping capacity to match the inflow hydrograph. A limited storage capacity must be provided in the pumping station wet well, but additional storage may serve to equalize the incoming flow and will reduce the required pumping capacity. The designer's objective is to establish a cost-effective balance between wet well storage volume and pumping capacity, and the possibility that the wet well will be partially full with limited equalization capacity should be considered as a worst-case scenario.

Future development of the watershed should be considered and incorporated into the derivation of the inflow hydrograph. If the watershed already is fully developed, and neither the impervious area nor the collection system will change significantly in the future, the inflow hydrograph for the pumping station may be based on existing conditions. However, if the watershed characteristics or collection system will be modified by future development, it is judicious to plan and design for both present and anticipated future conditions over the life of the pumping station.

One common and accepted way of providing increased future capacity is to leave space for additional pumps in the station. When this is done, the capacity of the station inflow and discharge piping system, wet well hydraulics, and power and control system also should be designed for future requirements. However, care must be exercised so as not to oversize the pump suction pipes to the point of causing a gross mismatch between them and the pump suction. Extreme mismatches between the diameters could lead to inadequate hydraulic operation of the smaller, initial-year pump. Chapter 5 presents details of pump suction piping.

2.2 Methods of Developing Inflow Hydrographs

There are several generally accepted methods for calculating the inflow hydrograph, including those developed by the U.S. Bureau of Reclamation (USBR), the Natural Resources Conservation Service (NRCS), and the U.S. Geological Survey (USGS). One method, still in wide use, is the rational method. Although several state and local agencies now prefer or require newer computerized methods of analysis, the rational method remains popular for analysis of smaller, less complex systems because of its simplicity.

2.2.1 Rational Method

The rational method of hydrologic analysis consists of an empirical formula that attempts to relate rainfall and runoff through use of a single coefficient.

The coefficient must account for all losses in the watershed, including infiltration, evapotranspiration, and surface depression storage. The rational method is discussed extensively in *Design and Construction of Sanitary and Storm Sewers* (ASCE & WEF, 1969).

There have been numerous efforts to improve the accuracy of estimates used for the equation's runoff coefficient, to the point that the use of the rational formula itself is often secondary. Moreover, because the rational method provides only an estimate of the peak runoff rate, additional methodology is applied to derive the runoff hydrograph.

The basic equation is

$$Q = C\,i\,A \tag{2.2}$$

Where

Q = the peak runoff rate, L/s (cu ft/sec);
C = the runoff coefficient reflecting characteristics of the drainage area and the collection system;
i = the average rainfall intensity, cm/h (in./hr); and
A = the drainage area, ha (ac).

The peak rate of runoff at any point is a direct function of the average rainfall intensity during the time of concentration to that same point. The frequency of the peak discharge is the same as the frequency of the average rainfall intensity. Finally, the time of concentration is the time required for the runoff to become established and flow from the most remote part of the drainage area to reach the point under design.

The last assumption applies to the part of the watershed most remote in time, not necessarily in distance. When using the rational method, average rainfall intensities have no time-sequence relationship to the actual rainfall pattern during the storm, and the intensity duration curve is not a time-sequence curve of actual precipitation. In addition, the rational method does not incorporate effects of preceding storm events and the way in which they may have altered the ground condition (such as saturation). Failure to recognize these points has led to improper use of the rational method in attempts to evaluate the effects of antecedent precipitation and infiltration.

Use of the rational method should be limited to small watersheds (fewer than approximately 80 ha [200 ac]) (Thomason, 2019). For larger watersheds, more reliable and accurate hydrograph development methods (such as those developed by USBR, NRCS, and USGS) are more appropriate. In these cases, a simple linear relationship with a single runoff coefficient cannot produce an adequate prediction of runoff, which involves storage, subsurface flow, stage discharge and storage, and phenomena previously discussed. Determination of the variables used in the rational method (Equation 2.2) are discussed in *Design and Construction of Sanitary and Storm Sewers* (ASCE & WEF, 1969). These variables, which also are used in other methods, include area, rainfall, and runoff.

The area tributary to any point under consideration in a storm collection system must be determined. This information should include the component parts of the drainage area, land use, soil characteristics, general slopes in the topography, and presence of drainage features.

Rainfall information includes intensity (i), duration, and frequency determination, all of which are interrelated. The average frequency of rainfall occurrence used for design determines the degree of protection afforded by a given storm sewer system. This protection should be consistent with the amount of damage prevented, and thus involves an economic analysis of cost (prevention) and benefit (protection). In practice, detailed cost-benefit studies typically are not conducted for small, urban, storm drainage projects. Judgment supported by records of performance in other similar areas is typically the basis for selecting the design frequency.

The ranges of frequencies used are as follows:

- For storm sewers in residential areas, 2 to 10 years, with 5-year frequencies being most common.
- For storm sewers in commercial and high-value districts, 10 to 50 years, depending on economic justification.
- For flood protection works, 50 years or more with 100 years common for many public works projects.

Note that in the latter two situations, a more detailed economic analysis frequently is justified because of the greater cost of providing additional protection.

Considerations that may lead to the use of other design storm frequencies include

- Potential for future relief from drainage problems in parts of the system (e.g., detention or flow equalization).
- Combined sewer systems that warrant using higher design rainfall frequencies.
- Special structures, such as pumping stations, that justify higher design storm frequencies.
- Lack of funding to provide the level of protection desired.
- Requirements of regulatory agencies (e.g., flood insurance program requirements and municipal, state, or federal regulations).

The cost of providing additional protection should be compared to the potential damage from flooding. For stormwater pumping stations, at least a brief economic analysis of a more conservative rainfall frequency design typically is justified. Figure 2.1 illustrates a typical economic analysis obtained by plotting both the costs because of damage from flooding or inadequate pumping capacity versus costs of providing increased pumping capacity. The sum of these two costs generally will yield a minimum point that indicates the minimum cost balance.

Basic rainfall data are derived from gauge measurements of rainfall. Length of record, gauge accuracy and placement, and density significantly affect the reliability of rainfall data.

It is important to note the relationship between total accumulated rainfall versus the length of the storm event and the return frequency (Frederick et al., 1977; Perica et al., 2018; U.S. Department of Commerce, 1956, 1960, 1961). Longer-duration storm events generally will have a greater total accumulation because the rainfall is accumulated over a longer period of time but may have a lower peak intensity (instantaneous intensity). Likewise, the longer the return period the greater the total accumulation. Rainfall intensity-duration curves for the selected locality and frequency may be prepared from these and similar charts. Figure 2.2 illustrates a curve of this type.

The relationship of rainfall intensity duration, storm pattern, time of concentration, and average intensity is discussed in *Design and Construction*

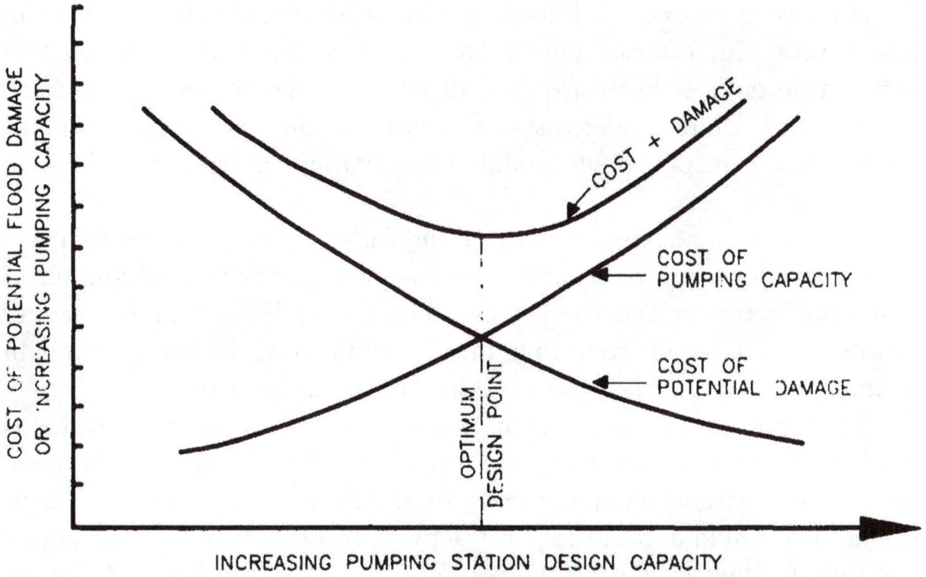

FIGURE 2.1 Typical Cost Optimization Curves Based on Potential Flood Damage Compared to Increased Pumping Capacity

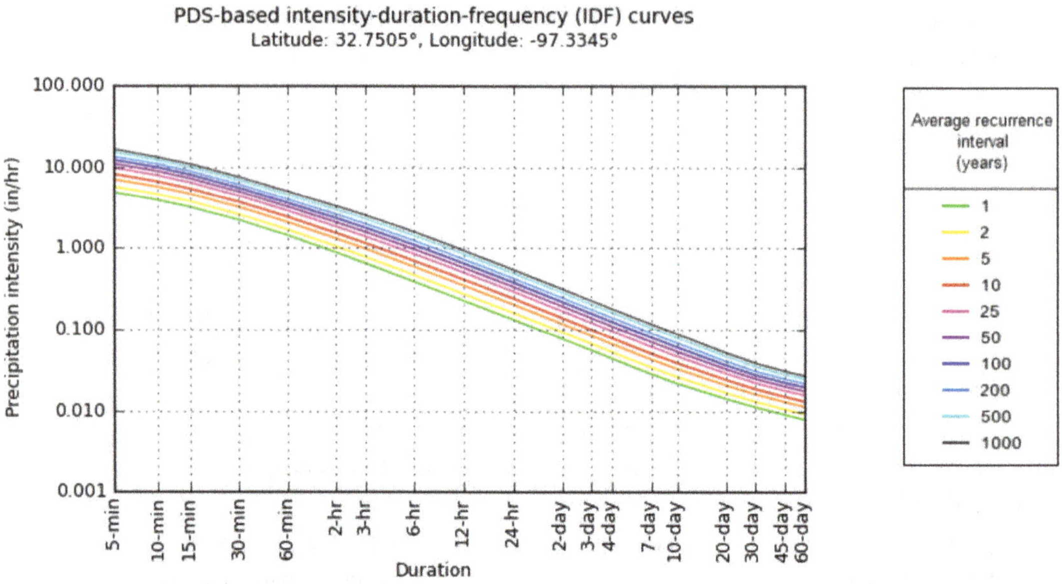

FIGURE 2.2 Intensity-Duration Rainfall Curves, Fort Worth, Texas

of Sanitary and Storm Sewers (ASCE & WEF, 1969). An estimate of the time of concentration to the point under consideration is made so that the average rainfall rate for the design period can be determined. For urban storm sewers, the time of concentration consists of the inlet time plus the time of flow in the sewer from the most remote inlet to the point under consideration.

Decreasing point rainfall intensity values for large areas may result in lower average intensities. Figure 2.3 (U.S. Department of Commerce, 1961) is based on data gathered from 20 dense networks of gauges in various regions of the United States east of the 90th meridian and gives factors for converting the average point rainfall intensity to the average intensity over the area of interest.

Care must be exercised when working with irregularly shaped drainage areas. Occasionally, a part of the area having a shorter time of concentration—and thereby subject to a higher-intensity rainfall—may cause a greater runoff rate at a design point than that contributed by the entire area with its longer time of concentration and lower rainfall intensity.

The runoff coefficient (C) is the variable of the rational method that is hardest to determine. Its use in the formula implies a fixed ratio between rainfall intensity and resultant runoff for any given drainage area. The coefficient also should account for abstractions or losses between the rainfall and runoff volumes, which may vary for a given drainage area and that are influenced by differing hydrological, climatological, and seasonal conditions.

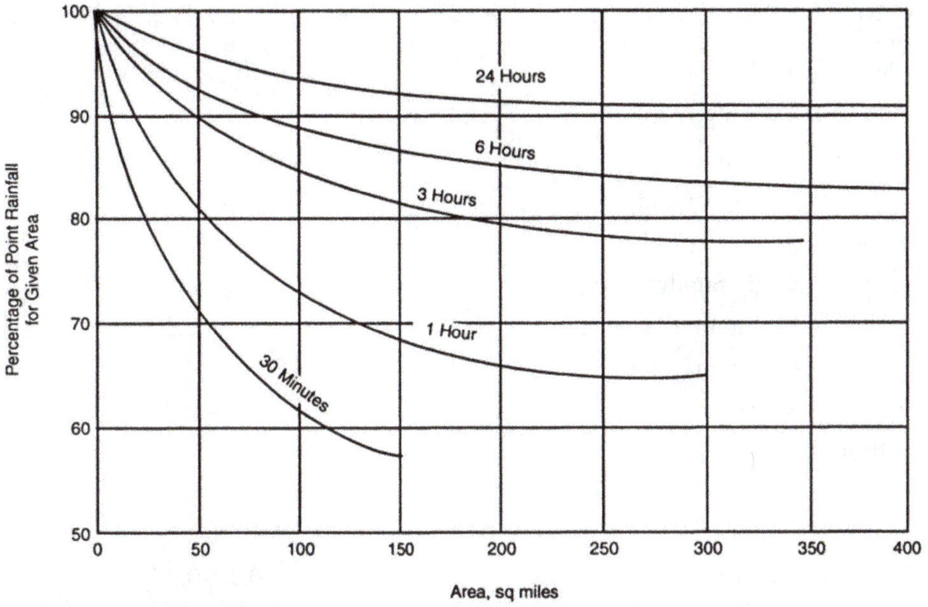

FIGURE 2.3 Converting Average Point Rainfall to the Average Over the Area of Interest (Area–Depth Curves) (sq mi × 2.59 = km^2) (U.S. Department of Commerce, 1961)

These losses include interception by vegetation, infiltration to permeable soils, retention in surface depressions and impoundments, and transpiration.

The use of average coefficients for various surface types, which are assumed not to vary through the duration of the storm, is common. The range of coefficients, clarified with respect to the general character of the tributary area reported in use, is given in Table 2.2.

Often, it is desirable to develop a composite runoff coefficient based on the percentage of various kinds of surface in the drainage area. This procedure frequently is applied to typical "sample" blocks as a guide to selecting reasonable values of the coefficient for an entire area. Table 2.3 shows coefficients reflecting surfaces in use. These coefficients assume that the design storm does not occur when the ground surface is frozen and impermeable, yet this condition existed during some flood disasters. Therefore, this assumption is a weakness of the rational method.

To apply the rational method to the design of stormwater pumping stations, it is necessary to develop an inflow hydrograph (Williams, 1950). By itself, the rational method produces only a peak flow that is assumed to be constant once the entire watershed is contributing to the flow (i.e., the time of rainfall is greater than the time of concentration). Several simplified methods have been proposed to develop an inflow hydrograph from

TABLE 2.2 Range of Runoff Coefficients for Various Tributary Areas

Description of Area	Runoff Coefficients
Business	
Downtown	0.70–0.95
Neighborhood	0.50–0.70
Residential	
Single-family	0.30–0.50
Multiunit, detached	0.40–0.60
Multiunit, attached	0.60–0.75
Apartment	0.50–0.70
Industrial	
Light	0.50–0.80
Heavy	0.60–0.90
Parks, cemeteries	0.10–0.25
Playgrounds	0.20–0.35
Railroad yard	0.20–0.35
Unimproved	0.10–0.30

TABLE 2.3 Range of Runoff Coefficients for Various Surface Types

Character of Surface	Runoff Coefficients
Pavements	
Asphaltic and concrete	0.70–0.95
Brick	0.70–0.85
Roofs	0.75–0.95
Lawns, sandy soil	
Flat, 2%	0.05–0.10
Average, 2% to 7%	0.10–0.15
Steep, 7%	0.15–0.20
Lawns, heavy soil	
Flat, 2%	0.13–0.17
Average, 2% to 7%	0.18–0.22
Steep, 7%	0.25–0.35

the peak flow produced by the rational method. One of the most common methods is the development of a triangular hydrograph, where the peak flow is the output from the rational method and the base of the hydrograph is made up of the time-to-peak (Tp) and the time-of-recession (Tr) (Figure 2.4) (U.S. Department of the Interior, 1965).

For most ungauged small watersheds, the Natural Resources Conservation Service has shown that the time for a storm event hydrograph to cease after peaking (Tr) is related to the time-to-peak (Tp) as follows:

$$Tr = 1.67Tp \tag{2.3}$$

Such simplified approaches to storm event analysis should be applied only to small watersheds of fewer than 80 ha (200 ac). Even then, any available information affecting runoff characteristics also should be considered. An attempt should be made to balance the excess rainfall with the volume of runoff water represented by the area under the triangular hydrograph. The designer should consider that runoff volume may be equally as important as peak flowrate and evaluate the possibility of a longer-duration storm that

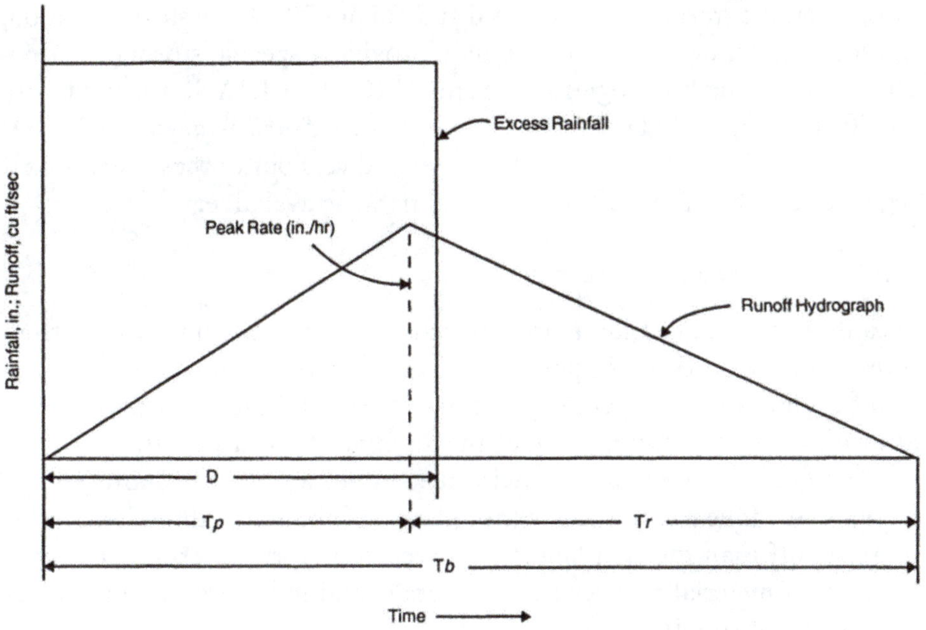

FIGURE 2.4 Triangular Hydrograph Analysis (U.S. Department of the Interior, 1965)

Note. Tp = time in hours from start of rise to peak rate; Tr = time in hours from peak rate to end of triangle; D = rainfall excess period, hours; and Tb = time base of hydrograph (cu ft/sec × 28.32 = L/s, in. × 25.4 = mm, and in./hr × 25.40 = mm/h)

might overwhelm the pumping station. By rearranging the rational method formula to solve for intensity given the maximum pump rate, the designer can better understand the risk of such a scenario and compare the intensity to the intensity–duration–frequency curve. If the coordinate duration is significantly longer than the *Tr* value assumed above, the designer should consider a more robust method.

2.2.2 Curve Number Method (Natural Resources Conservation Service Method)

The NRCS developed methods to be applied in estimating the runoff hydrograph for small urban watersheds (NRCS, 2020; U.S. Department of Agriculture [USDA], 1975). The following subsections address parameters involved in the curve number method.

2.2.2.1 Rainfall

Many rainfall hyetographs are available, but in keeping with a hypothetical design storm, it is advisable to use a balanced frequency storm methodology wherein the average rainfall intensity for a given duration is nested within each larger duration and centered at a specific proportion of the storm. In most cases, the storm will be centered at 50% or 67% of the storm duration, but local, state, or federal criteria may provide a specific storm centering. Other rainfall methodologies such as the NRCS Type I, IA, II, III, and IV are specified in Appendix D of *Urban Hydrology for Small Watersheds* (USDA, 1975). The NRCS has been producing revised versions of these curves on a regional level based on Atlas 14 rainfall data, as available.

2.2.2.2 Runoff Volume Parameters

Because urban areas seldom are covered completely by impervious structures, soil properties are important in estimating the total volume of direct runoff. Infiltration and percolation rates of soils indicate their potential to absorb rainfall and thereby reduce the amount of direct runoff.

The type of cover and its hydrologic condition affect runoff volume through its influence on the infiltration rate of the soil. Fallow land yields more runoff than forested land for a given soil type. Covering areas with impervious material reduces surface storage and infiltration and increases the volume of runoff.

Some rainfall is retained on the surface and by vegetation before runoff begins. Interception is rainfall that is caught (e.g., by foliage, twigs, branches, and leaves). This rainfall is lost to evaporation and never reaches the ground surface. Increasing the vegetal cover increases interception.

Surface depression storage begins when precipitation exceeds infiltration. Overland flow starts when surface depressions are full. The water captured in depression storage is not available as direct runoff.

Initial abstraction, which is the sum of interception, depression storage, and infiltration before runoff begins, occurs on all types of cover, from pasture in good condition to concrete pavement. The amount of initial abstraction is less on concrete pavement than on pasture.

2.2.2.3 Time Parameters

The slopes of storm sewers, street gutters, roads, overland flow areas, and stream channels are significant in determining travel times through urban watersheds.

The flow length is the distance the runoff must travel from the most distant portion of the watershed to the pumping station. This is not necessarily a straight line because flow must follow the topography.

The degree of surface roughness influences the time of runoff. Rough surfaces include woodland, grassland, and natural channels. Smooth surfaces include paved surfaces and concrete-lined channels.

Following is a brief outline of the methodology used in the NRCS urban runoff method (USDA, 1975), which is recommended for drainage areas up to 52 km^2 (20 sq mi). Peak discharge and total direct runoff techniques may be used for areas up to 800 ha (2000 ac) (USDA, 1975).

2.2.2.4 Determination of Runoff Volume

Runoff volume is derived from information on the rainfall amount and watershed characteristics. A combination of a hydrologic soil group, land use, and treatment class is used to determine the hydrologic soil-cover complex. The effect of the hydrologic soil-cover complex on the volume of rainfall runoff is represented by a runoff curve number. If the rainfall amount and curve number are known, the volume of runoff from the watershed can be estimated.

2.2.2.5 Time of Concentration, Travel Time, and Lag

As with the rational method, "time of concentration" is the time it takes for runoff to travel from the hydraulically most distant part of the watershed to the point of reference and is computed by determining the water travel time through the watershed. In hydrograph analysis, it is the time from the end of excessive rainfall to the point of inflection on the falling limb of the hydrograph. Lag can be considered a weighted time of concentration and is related to the physical properties of a watershed, such as area, length, and slope. In simple hydrograph analysis, *lag* is the time from the center of mass of an excessive rainfall to the peak rate of runoff. The time of concentration determines the shape of the runoff hydrograph. Thus, changes in the time of concentration cause changes in the resulting hydrograph. The extent of urbanization and stream modification affects the travel time of water through the watershed. Travel time, in turn, changes the time of concentration.

Travel time and time of concentration can be determined using available tables and equations for overland flow, channel flow, and storm sewer flow. Lag time typically is calculated as a proportion (typically 0.6) of the time of concentration.

2.2.2.6 Peak Discharge

After the time of concentration and travel time, the drainage area, the runoff curve number, the 24-hour rainfall and its distribution, and the runoff in inches are known, the peak discharge can be obtained from NRCS tables (USDA, 1975).

2.2.3 Computerized Runoff Models

Computer models have become a key component of standard practice to analyze urban runoff quantity and quality and to enable a more thorough analysis of a given pumping station configuration. Most of these models are available in versions compatible with personal computers and have been adapted for user-friendly interface, easy data input, and graphical presentation of results.

Computer models are used to study quantity and quality problems associated with urban stormwater runoff in which three broad system objectives can be identified: planning, design, and operation. Typically, each objective will require a model with different, yet overlapping, characteristics. It should be noted that all models are based on a variety of assumptions, and the user must understand their nature and proper use to avoid erroneous model results.

2.2.3.1 Planning Models

Planning models can provide an overall assessment of the urban runoff problem and an estimate of the effectiveness and cost of abatement procedures. They are often used for an initial analysis of the rainfall-runoff relationship and may illustrate tradeoffs among various control options (e.g., treatment versus storage). These models have relatively long time steps (hours) and long simulation times (months or years). Data requirements typically are minimal, and their mathematical complexity is low. A good example is the Hydrologic Modeling System developed by the Corps of Engineers Hydrologic Engineering Center (Bartles et al., 2021), which can simulate complex hydrographs and basic pump operations.

2.2.3.2 Design Models

Design models offer the user a more detailed simulation of a single storm event. Most of the current design models provide a complete description of flow and pollutant routing from the area of rainfall through the entire collection system, and often to the receiving waters as well. Design models can be used to predict flows and concentrations anywhere in the rainfall and collection system and can illustrate more clearly how abatement procedures or design options affect them. These models are a rapid and cost-effective means of determining least-cost abatement procedures for both quantity and quality problems in urban areas.

Design models are used to simulate a single storm event and have short time steps (seconds) and short simulation times (minutes). Data requirements may be moderate to extensive, depending on the model used, and mathematical analysis is more complex than at the planning level. The most complex models may include two-dimensional analysis of surcharged flows that allow the designer to evaluate damages associated with the lack of pumping or storage capacity indicated above. Moreover, evaluation of upland drainage system limitations may change the inflow hydrograph by allowing more accurate accounting of surface storage.

The U.S. EPA Storm Water Management Model (SWMM) is an example of a design-level model developed to simulate urban stormwater quantity and quality processes and to assess the effectiveness of alternative abatement and mitigation procedures (Rossman & Simon, 2022). Private software developers have provided a user-friendly graphical interface and more powerful tools for interpreting model results with such software suites as PC-SWMM, XP-SWMM, and InfoWorks ICM, among others.

2.2.3.3 Operational Models

Operational models are used to either produce actual control decisions or to recommend a control strategy to operating personnel during a real storm event. For real-time operational models, rainfall data are telemetered from remote stations and entered directly to the model, which then is used to predict collection system responses a short time into the future. Various control options can then be used (e.g., in-system storage, diversions, or regulator settings).

Operational models are developed from existing sophisticated design models that have been adapted and calibrated for a particular system. Examples are those designed for Minneapolis–St. Paul (Bowers et al., 1968), Seattle (Leiser, 1974), and the Washington Suburban Sanitary Commission (Washington Suburban Sanitation Commission, 1983).

2.2.3.4 Sensitivity Analysis

Due to the uncertainty inherent in hydrologic parameters, it is good practice to evaluate the stormwater pumping station under a variety of conditions within its tributary area, including all input variables to the determination of the inflow hydrograph. This is especially true for complex stations that have a significant volume within the wet well/sump and stations with multiple pumps that are operated at various levels. The sensitivity analysis may reveal inefficiencies in the setpoints of the different pumps and provide the designer more insight to the operations such that a more economical design may be achieved than that suggested by the single design case. Computer models can make this a highly automated process more dependent on managing data inputs and outputs than concerted design efforts.

3.0 COMBINED FLOW SYSTEMS

Combined sewer systems are those collection, conveyance, and pumping systems designed to carry both sanitary flow and stormwater flow in a single system. Although few combined sewer systems now are being constructed, many cities have extensive existing combined sewer systems dating from several years ago. For reasons including cost, politics, and feasibility, it is unlikely these systems will be modified to accommodate segregated flows in the near future. For that reason—and because the engineer often must expand or modify a combined sewer system—a brief description of the methods used to determine combined flows is presented.

The design of combined sewers is the same as for storm sewers, though the design of pumping stations for combined flows must consider wastewater flows and volumes. Long detention times in a wet well cause septic conditions and odor problems. Thus, to determine the capacity for a combined flow pumping station, both sanitary wastewater and stormwater flows must be considered. For low flow purposes (e.g., dry weather conditions), wastewater rates are considered alone. Pumping capacity and wet well volume should be designed to provide adequate but short detention times for wastewater. For higher, wet weather flows, the sanitary wastewater flow can be added to the stormwater flow (or neglected altogether if insignificant), and additional pumps in the same station can be sized to handle these large stormwater flows.

Bypass and overflow devices on the collection system should be considered when determining system flows and designing pumping station capacities. Design examples illustrating the calculation of stormwater flows, wastewater flows, and combined flows are presented in *Wastewater Engineering* (Metcalf & Eddy, 2014).

4.0 FLOW REGULATION

To a limited degree, the pumping station wet well volume may be used to equalize inflow to the pumps, and thus reduce the size or number of pumps required. Although, theoretically, this storage can be used for both stormwater and wastewater pumping stations, long storage periods in wastewater pumping stations could be impractical because of the problems of solids settlement and odor production.

To size the wet well for a stormwater pumping station, the inflow hydrograph must be routed through the wet well. The general equation for routing a flow through any storage facility is

$$\text{Rate of inflow} - \text{rate of outflow} = \text{rate of change of storage} \quad (2.4)$$

The method of reservoir routing is covered in most hydrologic texts (Chow, 1964; Commonwealth of Pennsylvania Department of Environmental Protection, 1997; Linsley et al., 1958). In a pumping station, the rate of inflow is derived from the inflow hydrograph, and the rate of outflow is the pumping rate. The rate of change of storage is a direct function of the shape of the entrance to the pumping station and wet well. The piping system leading to the pumping station may also contribute to storage, depending on its invert elevation relative to that of the wet well. When available, this latter storage may be significant.

The total pump capacity must be balanced with the storage volume provided to obtain the most economical balance between the two. The method of mass curves also can be applied to the graphical solution of the routing equation (Linsley & Franzini, 1972).

5.0 QUALITY OF INFLUENT

The quality characteristics of the influent are as important as the quantity and type of influent to the pumping station. These quality characteristics will affect the overall design of the proposed facilities, influence the selection of the equipment, and set certain required operational procedures once constructed. The following types of characteristics should either be determined from existing information or a reasonable judgment should be made of their likely presence in the influent (Jones & Sanks, 2008):

- Pumps should be sized to accommodate the largest solids, if possible. In the case of storm or combined flows, it may be necessary to add mechanical screening or grinding to the facility's design. It is common practice to protect wastewater pumps from solids and rags by installing comminutors or grinders in influent channels before the wet well. These additional items will increase facility costs and have a significant effect on O&M costs.
- Unusual grit quantities could determine special materials for pump construction or the need for upstream grit removal.
- Corrosive constituents, particularly when dealing with industrial waste flows.
- The presence of toxic, explosive, or hazardous materials may establish requirements for monitoring and special venting.
- Possibility of septic wastewater and large quantities of hydrogen sulfide or other odor-causing substances. These characteristics may require the use of special coatings to reduce corrosion, venting, and odor control for operator safety and of variable-speed pumping equipment to reduce residence wet well time.

6.0 REFERENCES

American Society of Civil Engineers, Environmental and Water Resources Institute, & Water Environment Federation. (2007). *Gravity sanitary sewer design and construction* (2nd ed.; ASCE Manuals and Reports on Engineering Practice No. 60; WEF Manual of Practice No. FD-5). American Society of Civil Engineers.

American Society of Civil Engineers & Water Pollution Control Federation. (1969). *Design and construction of sanitary and storm sewers*. American Society of Civil Engineers and Water Environment Federation.

Bartles, M., Brauer, T., Ho, D., Fleming, M., Karlovits, G., Pak, J., Van, N., & Willis, J. (2021). *Hydrologic modeling system HEC-HMS user's manual*. U.S. Army Corps of Engineers Hydrologic Engineering Center.

Bowers, C. E., Harris, G. S., & Pabst, A. F. (1968). *The real time computation of runoff and storm flow in the Minneapolis–St. Paul interceptor sewers* (Memo No. M-118). University of Minnesota, Saint Anthony Falls Hydraulic Laboratory.

Chow, V. T. (1964). *Handbook of applied hydrology.* McGraw-Hill.

Chow, V. T., Maidment, D., & Mays, L. (1988). *Applied hydrology.* McGraw-Hill.

Commonwealth of Pennsylvania Department of Environmental Protection. (1997, October). *Domestic wastewater facilities manual: A guide for the preparation of applications, reports and plans.* Bureau of Water Quality Protection.

Fair, G. M., & Geyer, J. C. (1954). *Water supply and waste-water disposal* (1st ed.). John Wiley & Sons.

Frederick, R. H., Myers, V. H., & Auciello, E. P. (1977). *Five-to-60-minute precipitation frequency for the eastern and central United States* (NOAA Technical Memo, NWS HYDRO-35). National Oceanic and Atmospheric Administration, National Weather Service, Office of Hydrology.

Jones, G. M., & Sanks, R. L. (2008). *Pumping station design* (3rd ed.). Butterworth-Heinemann.

Leiser, C. P. (1974). *Computer management of a combined sewer system* (EPA-670/2-74-022). U.S. Environmental Protection Agency.

Linsley, R. K., & Franzini, J. B. (1972). *Water resources engineering.* McGraw-Hill.

Linsley, R. K., Kohler, M. A., & Paulhus, J. L. H. (1958). *Hydrology for engineers.* McGraw-Hill.

McJunkin, F. E. (1964). Population porecasting by sanitary engineers. *Journal of the Sanitary Engineering Division, Proceedings of the American Society of Civil Engineers, 90*(SA4).

McLean, J. E. (1952). More accurate population estimate by means of logistic curves. *Civil Engineering,* 22.

Metcalf & Eddy. (2014). *Wastewater engineering: Treatment and resource recovery* (5th ed.). McGraw-Hill.

Natural Resources Conservation Service. (2020). *National engineering handbook, part 630, hydrology.* U.S. Department of Agriculture, Natural Resources Conservation Service.

Pennsylvania Department of Environmental Protection. (1983). *Standards for sewage disposal facilities.*

Perica, S., Pavlovic, S., St. Laurent, M., Trypaluk, C., Unruh, D., & Wilhite, O. (2018). *NOAA atlas 14 precipitation-frequency atlas of the United States.* Volume 11, version 2.0: Texas. National Oceanic and Atmospheric Administration, National Weather Service.

Rossman, L. A., & Simon, M. (2022). *Storm water management model user's manual version 5.2*. U.S. Environmental Protection Agency.

Thomason, C. (2019). *Hydraulic design manual*. Texas Department of Transportation.

United Nations. (n.d.). *Population projections by the fertility rate method*. United Nations Publication.

U.S. Department of Agriculture. (1975). *Urban hydrology for small watersheds* (Technical Release No. 55). U.S. Department of Agriculture, Soil Conservation Services, Engineering Division.

U.S. Department of Commerce. (1956). *Rainfall intensities for local drainage design in western United States, for durations of 20-minute to 24-hour and 1 to 100-year return periods* (Technical Paper No. 28). U.S. Department of Commerce, Weather Bureau.

U.S. Department of Commerce. (1960). *Generalized estimates of probable maximum precipitation for the United States west of the 105th meridian for areas to 400 square miles and durations to 24 hours* (Technical Paper No. 38). U.S. Department of Commerce, Weather Bureau.

U.S. Department of Commerce. (1961). *Rainfall frequency atlas of the United States for durations from 30 minutes to 24 hours and return periods from 1 to 100 years* (Technical Paper No. 40). U.S. Department of Commerce, Weather Bureau.

U.S. Department of the Interior. (1965). *Design of small dams*. U.S. Department of the Interior.

Washington Suburban Sanitation Commission. (1983). Sewer system computer model: Design report, volume I, II, III, and IV. Unpublished engineering report. Washington Suburban Sanitation Commission.

Water Environment Federation & American Society of Civil Engineers. (2018). *Design of water resource recovery facilities* (6th ed.; WEF Manual of Practice No. 8/ASCE Manuals and Reports on Engineering Practice No. 76). Water Environment Federation.

Water Environment Federation, American Society of Civil Engineers Environmental, & Water Resources Institute. (2009). *Existing sewer evaluation and rehabilitation* (WEF Manual of Practice No. FD-6/ASCE/EWRI Manuals and Reports on Engineering Practice No. 62). Water Environment Federation.

Williams, G. R. (1950). Hydrology. In *Engineering Hydraulics*. John Wiley & Sons, p. 309.

3

Station Configuration and Design

This chapter discusses the configuration of various sizes and types of pumping stations and factors influencing the design of pumping stations, which include the needs of the end user and satisfying environmental, social, and political constraints.

Station capacity, as described in Chapter 2, establishes the range of flows to be pumped, from which the number and size of pumps can be determined. Before detailed plans and specifications are prepared, a preliminary design layout should be presented to end users with scaled sketches showing the number of pumps; pump type, selection, and arrangement; provisions for appurtenances; and site dimensions. Station configuration also can be affected by fluid characteristics; site constraints such as accessibility, maintenance location, and zoning ordinances; and appurtenances such as bar screens or comminutors, which may be needed to keep large objects from flowing down the sewer into the pumping station wet well. The pumping station and force main should be designed as a unit.

Because technical, environmental, social, and political constraints are unique to each facility, not all pumping stations are identical. This chapter provides guidance for the design of facilities to satisfy varying needs. Multiple design disciplines including civil, structural, mechanical, electrical engineering, and architectural and instrumentation are required for successful design.

1.0 LOCATION CONSIDERATIONS

In determining the need for a wastewater or stormwater pumping station, the hydraulics of a sewer or drainage collection system must be evaluated. If gravity flows can be achieved economically, as determined by comparing the cost of constructing the pumping station and annual operations and maintenance (O&M) expenses to the construction cost of gravity sewers or storm drains, the station should be eliminated. Location and size of a pumping station should be determined after an analysis of technical needs, environmental constraints, local zoning ordinances and restrictions, economics, and public input during the planning stage.

The preliminary layout of a station depends on the design capacity, type of station, number, type, and the immediate and future capacity of pumps. The need and flexibility for future expansion should also be evaluated. If the station has a wide range of flows, the provision for a second future force main will affect station configuration but offers a wider range of pump combinations and selections. Other considerations in locating a pumping station include land ownership, accessibility, site and surrounding conditions such as flood potential and drainage, and availability of utilities (water, electricity, natural gas, and telephone). Also, subsoil and groundwater conditions and the necessary station depth of the specific substructure must be studied during site selection.

Site design should consider pathways of radio telemetry with central or relay stations (for alarm signals). Alternatively, fiber optic lines provide a resilient method for communication with operations staff. If possible, stormwater pumping stations should be situated where water can be impounded with minimal potential for flood damage should inflow exceed station capacity. Impounding a portion of the stormwater inflow will control the rate of stormwater flow, thereby reducing the overall pumping capacity required for any given storm.

The selected station site should consider construction stage requirements, interceptor sewer and force main alignments, environmental effects, vehicular access, site grading, future removal and replacement of equipment, future expansion needs, noise constraints, and architectural and aesthetic needs. Malfunction of a wastewater or stormwater pumping station can cause considerable public nuisance and property damage; therefore, reliability of power supply (including standby power) and control systems in addition to selection of appropriate pumping equipment is essential. Chapters 4, 6, and 7 discuss the available types of drives and controls for pumping equipment and reliability of power supplies.

For small- to medium-capacity pumping stations, yard piping should be designed with bypass valving such that during total station failures, a transportable pumping unit such as engine-driven pumps can be brought in to provide emergency pumping. For larger pumping stations, use of transportable pumping units during emergencies may not be possible, so station reliability should be improved by installing standby pumps and generators. For rehabilitating existing pumping stations without standby power, this yard piping feature could be used. For new pumping stations, standby power is essential because of severe potential administrative fines against wastewater spillage. Stormwater pumping stations are required to operate during extreme weather conditions when power outages are likely and, therefore, standby power in the form of a generator or an alternative pump power drive unit such as a diesel or natural gas engine are necessary.

Because of the location of gravity sewers and drainage canals, many pumping stations are located within floodplains, floodway fringes, and floodways, and thus are prone to flooding. To protect pumping station structures and electrical and mechanical equipment from flood damage, site the operating floor (where electrical and control panels typically are located) above the base flood elevation (BFE) or design the pumping station against flooding. Although states or cities have different requirements, the BFE is considered the elevation 0.3 to 0.6 m (1 to 2 ft) above the maximum level floodplain. Pumping stations prone to dry weather flooding or storm surge with wave action should be raised a minimum of 0.6 m (2 ft) above the BFE. Where practical, critical equipment that must be replaced when flooded, such as electrical boxes and control panels, should be raised above the 500-year flood elevation. All entrances and non-watertight openings in the station should be at elevations above the maximum flood level. Also, it is recommended that pumping stations be designed to remain fully operational and accessible during periodic high flood conditions. These criteria should be verified with local regulations and ordinances. For streams with no delineated maximum-level floodplains, the design engineer should use appropriate computer models, such as the Hydrologic Engineering Center River Analysis System program developed by the U.S. Army Corps of Engineers (USACE), to determine the approximate maximum-level floodplain elevation.

2.0 STATION TYPES

The type of pumping station selected should be dictated by such factors as design capacity, function, type of pumps needed, cost, and environmental and aesthetic considerations. The equipment components needed in a particular station to satisfy function and design requirements (such as screening, odor control, standby power, monorail, and hoist) may determine the station configuration and type. While design pumping capacity is the most significant variable in selecting the type of pumping station, size of site and site limitations may also affect the type of station chosen. The following list identifies some of the factors that are influenced by site and flow constraints:

- Construction type
  - Prefabricated with package pumping equipment and controls
  - Custom built or built-in-place
  - Packaged equipment for installation in manholes or built-in-place underground structures
- Pump location
  - Inside wet well
  - Inside dry well
  - Above ground
- Motor and drive location
  - Above ground inside super structure
  - Below ground level inside dry well
  - Below ground level inside wet well (integral with most submersible pumps)
  - Enclosed indoors
  - Exposed outdoors
- Drive type
  - Electric motor
  - Diesel engine
  - Natural gas engine
  - Hydraulic power unit

2.1 Ranges of Flow

Compared with most wastewater pumping stations, stormwater stations typically handle larger flows, the largest of which are outside the scope of this design. Design recommendations for these larger stormwater stations can be found in the USACE's (2020) *Mechanical and Electrical Design of Pumping Stations*. To classify stations for handling either wastewater or smaller stormwater stations, ranges of flow can be used to distinguish several station types:

- Small flow stations: 30 L/s (500 gpm)
- Small to Medium flow stations: 30 to 200 L/s (500 to 3200 gpm)
- Medium to Large flow stations: 200 to 1260 L/s (3200 to 20 000 gpm)
- Large flow stations: 1260 to 12 620 L/s (20 000 to 200 000 gpm)

Note, stations larger than 12 620 L/s (200 000 gpm) require special considerations and are outside the scope of this manual.

Small flow stations can handle from a few houses to 500 or more residential buildings. Prefabricated package pumping stations typically should be considered for such applications, as should built-in-place stations with package pumping equipment. A pump may be permitted, but duplex pumping units are recommended for system reliability, with the second pump providing 100% standby capacity in the event of lead-pump failure. State and local requirements may dictate the redundancy for wastewater and drainage pumping stations and should be consulted before completing design. For a stormwater station that is expected to operate only during extreme events, a redundant pump would not be required as the owner will need all pumping capacity to handle such an event.

Figure 3.1 illustrates general configurations of package pumping stations. It shows duplex submersible pumping units in a single wet well sump, which is ideal for handling wastewater or stormwater for a single building or a group of buildings where blockage by large objects is unlikely.

Comminutors have been used to reduce the size of solids in the flow before passing through the downstream pump. If larger solids in the flow stream necessitate a comminutor, then a combination grinder pump and bar screen should be considered.

For the small stations, non-clog-type pumps capable of passing 65- to 80-mm (2.5- to 3-in.) solids or grinder pumps are required. Screening devices to prevent clogging of pumps are warranted if the selected pump cannot pass the minimum expected sphere size or the debris is irregular (e.g., household debris, tree branches, vegetation, construction materials, etc.), as can be anticipated with stormwater. The duplex pumps should be designed to handle 100% of peak design flow, with the second pump serving as a full standby unit.

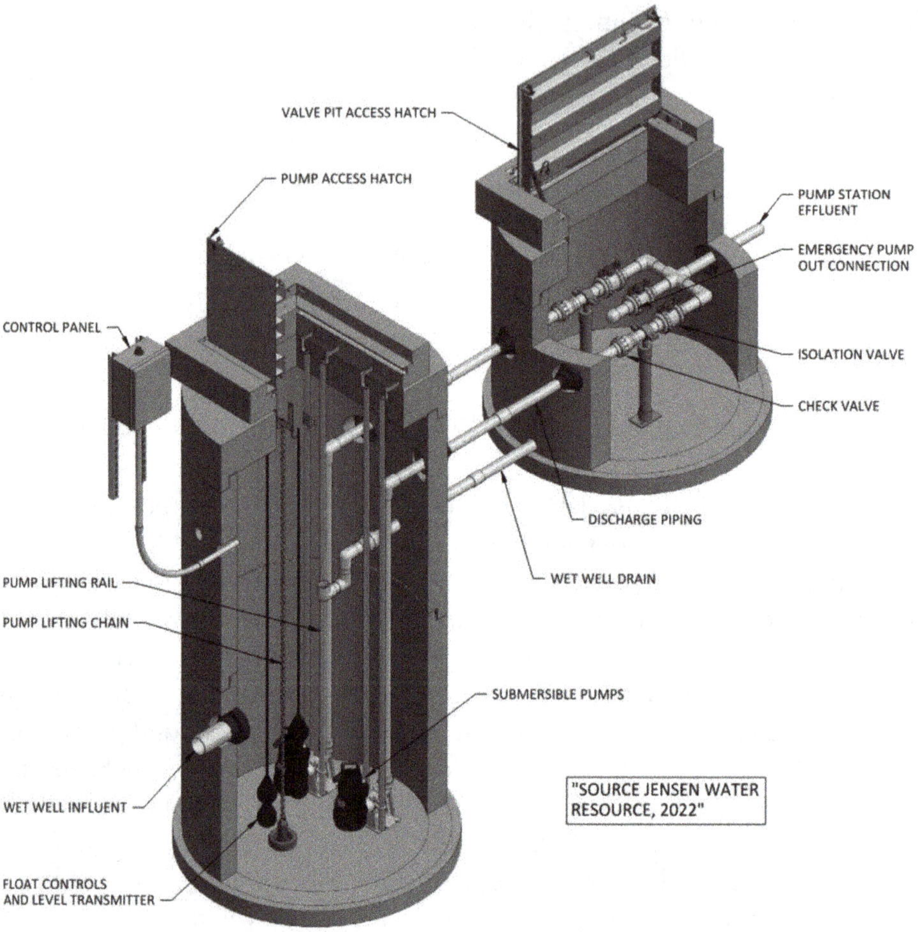

FIGURE 3.1 Package Submersible Pumping Station Configuration

Small-to-medium stations handling 30 to 200 L/s (500 to 3200 gpm) typically incorporate a minimum of two pumps, each having the capacity to match the maximum design flow, with the second pump providing 100% standby capacity. As the design capacity and pressure head requirements increase, a three-pump station should be studied, with the third pump providing 50% standby capacity. The three-pump station should be designed so that the station can pump the peak flow with the largest pump out of service. The three-pump station design provides added flexibility in meeting flow requirements. During an interim expansion stage, pumping capacity can be increased by replacing the impeller with the larger one, provided higher horsepower requirements are planned in the initial motor, motor starter, and electrical designs.

Though not to exclude prefabricated stations, the design engineer must consider access for maintenance and available built-in piping flexibility for future pump replacement or station expansion. Self-priming pumps should be limited to practical maximum suction lifts of approximately 6 m (20 ft). Figure 3.2 illustrates an example of a self-priming station for small-to-medium

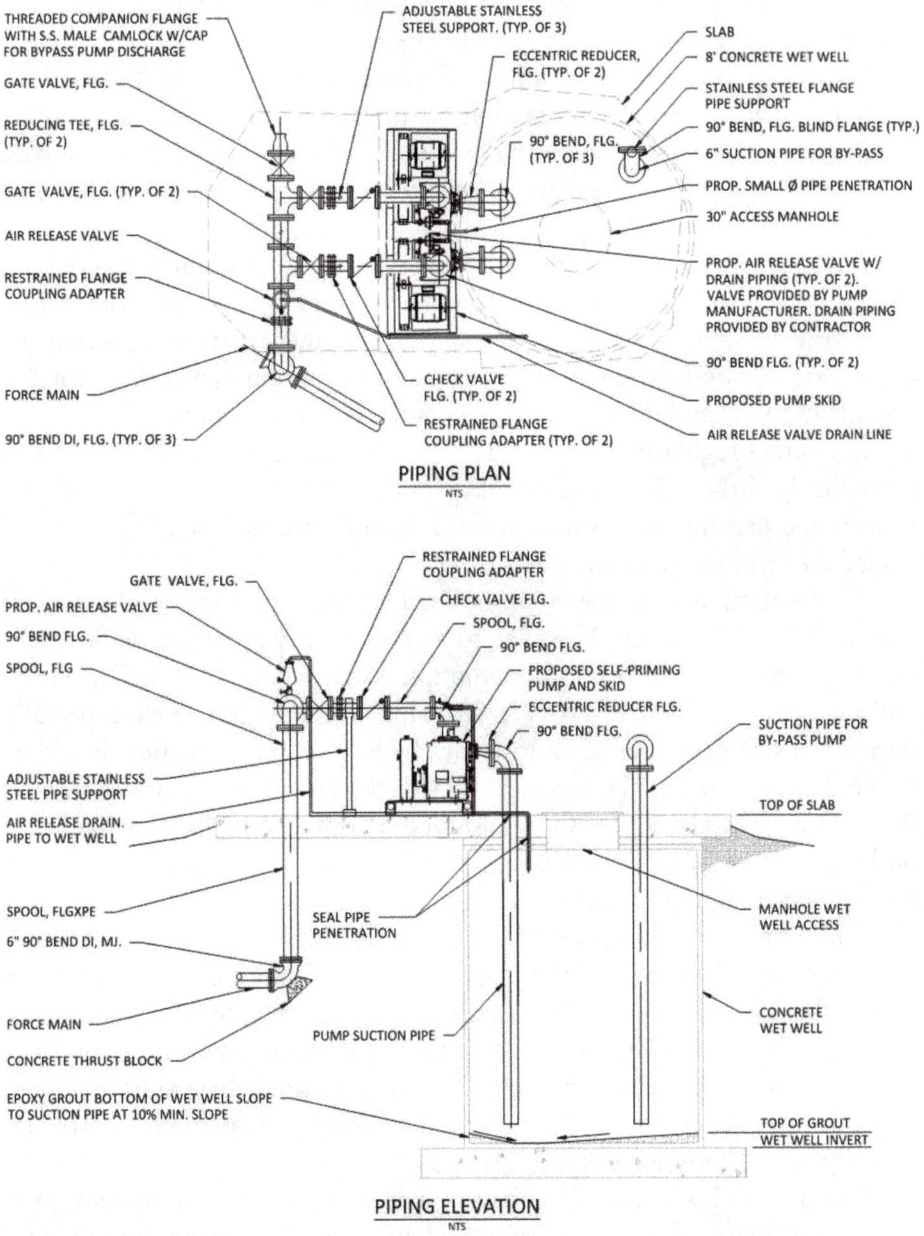

FIGURE 3.2 Small-to-Medium Pumping Station Configuration (Self-Priming Pumps)

capacity. Use of self-priming pumps should be evaluated carefully because the need for a priming device adds another level of mechanical maintenance.

To reduce electrical power surges for direct across-the-line starters for higher horsepower motors, soft-start motor controllers and variable-speed drives (typically of the variable-frequency type) have applications in this flow range. The additional electrical panels required for such equipment may affect the station building size and arrangement. These are discussed further in Chapter 6.

Figure 3.3 illustrates an example of a built-in-place station for small-to-medium capacity. The discharge piping and valve arrangements are shown. Owner preference may dictate the location of discharge piping and valves, although access for maintenance often dictates that valves be located above ground for ease of access.

Air diffusers in the wet well may be considered as an optional design. They are used to freshen the flow stream (for odor control downstream, in lieu of the air injection to the force main sometimes used) and prevent the deposition of solids. Because aerating the flow stream can release volatile organic carbons and odor-causing agents to the air, to comply with air pollution control regulations the inclusion of air scrubbers should be studied, particularly if there is an industrial component in the flow stream. If preventing the deposition of solids is the only concern, the use of shaped and sloped wet wells is recommended.

A two-pump arrangement can be used in larger stations handling more than 200 L/s (3200 gpm); however, to deal with varying ranges of flow efficiently often requires three or more pumps. A combination of smaller pumps and one large pump will provide flexibility for low flowrates, especially during the startup stage when flows can be low. Pump capacities should be selected to convey the maximum flow with the largest pump out of service. If variable-speed drives are to be installed, the drives might be sized based on larger future pumps (or a larger impeller) to save the cost of replacing them when pumping capacity is increased.

While prefabricated stations function adequately in certain medium-flow-range applications, a built-in-place station is the norm and provides greater flexibility and better satisfies site constraints and O&M requirements.

Figure 3.4 shows the plan of a station with a three-pump design (two existing pumps, one future pump). The entire station's hydraulic structure was built during the first phase to reduce overall construction costs and simplify expansion in the second phase.

The use of 3D modeling software has enhanced the designer's ability to coordinate between disciplines, identify conflicts during design, and better convey design intent to the construction contractor. Figure 3.4 also shows an isometric view of this submersible lift station built using 3D modeling software.

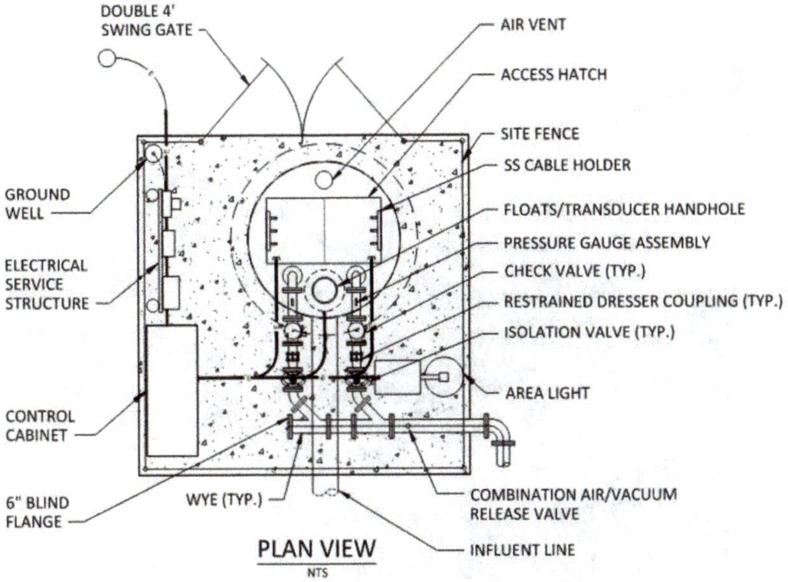

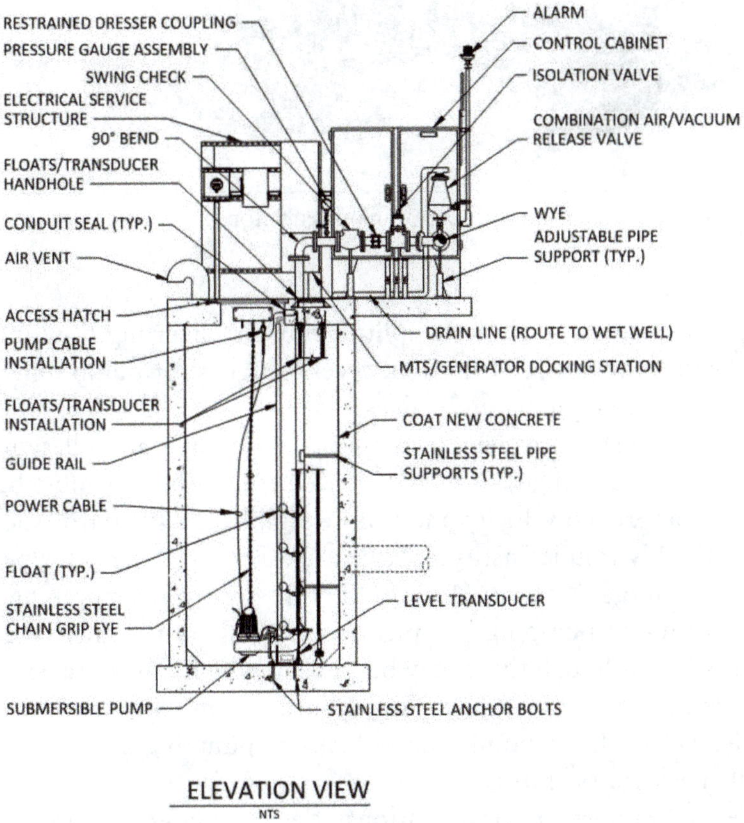

FIGURE 3.3 Small-to-Medium Pumping Station Configuration (Submersible Pumps)

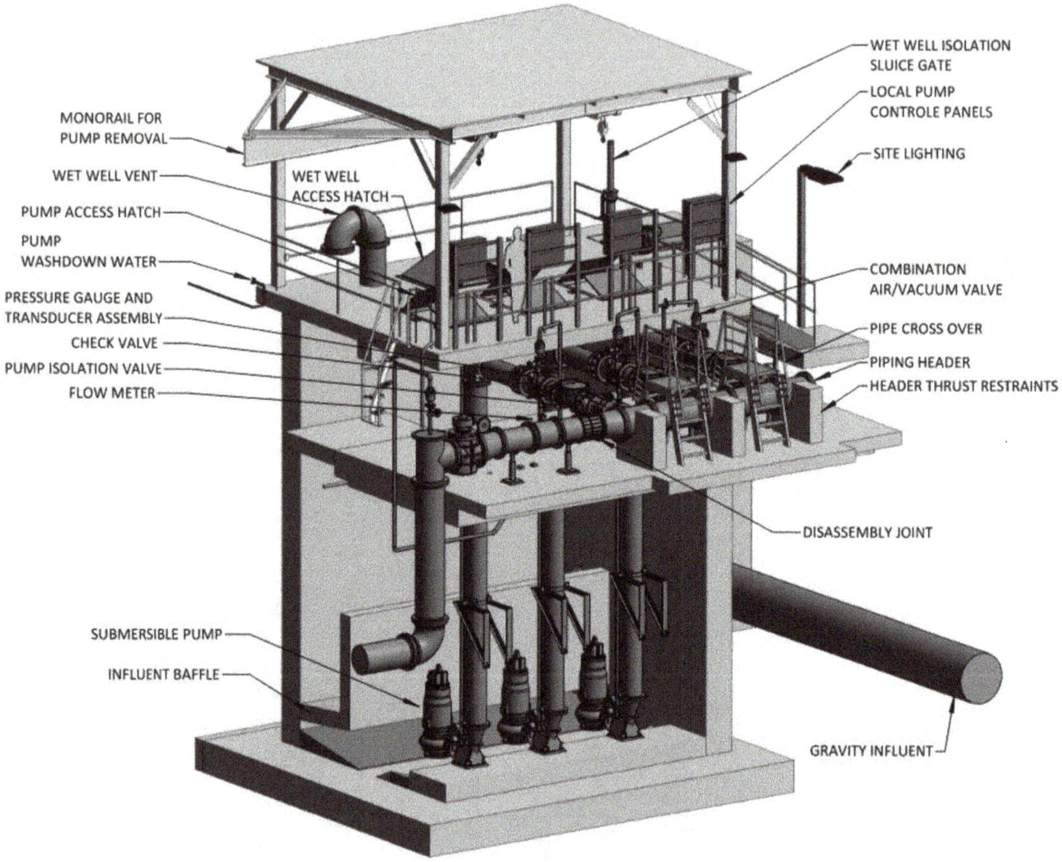

FIGURE 3.4 Medium-to-Large Pumping Station Configuration

The lift station includes a splitter box for diverting flow into the wet well, baffles for flow distribution, access hatches for pump removal, hoist, and all discharge piping and valves.

Figure 3.5 shows a general layout of a pumping station designed to handle 8760 L/s (200 mgd) of flow. Wet well configurations must be designed based on approach velocity and follow established standards such as those set by the Hydraulic Institute. Figure 3.6 shows a wet-pit-type pumping station for large (7.5-m^3/s [120 000-gpm]) stormwater pumping flows.

Stormwater pumping is typically used only when gravity drainage is not feasible, although there may be cases in which it is more economical to pump than to construct gravity conveyance systems. Storage in storm drains, canals, wet wells, or ponds can reduce the pumping capacity required to handle peak runoff rates.

Several important considerations affect planning and site selection for pumping stations. Access necessary for safe operation, maintenance, and

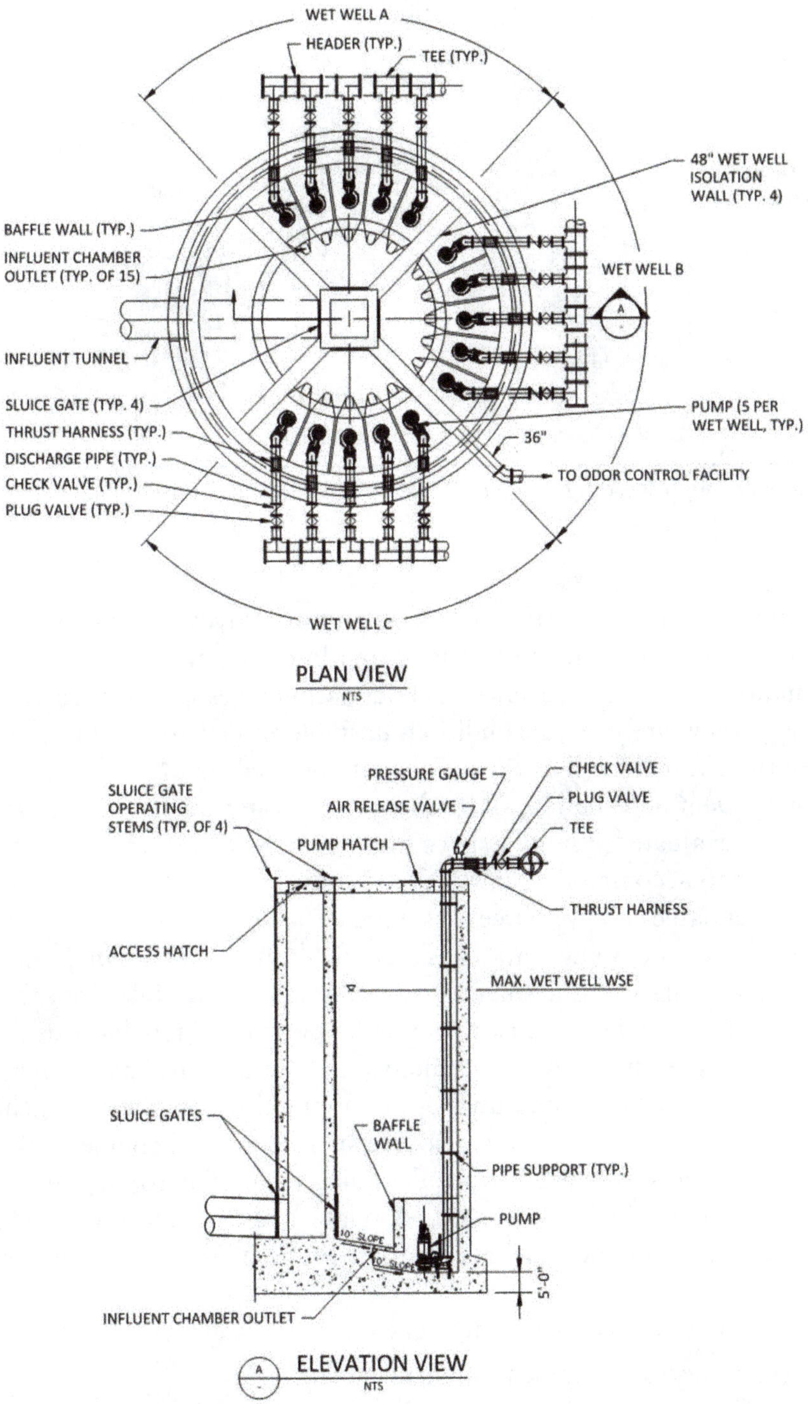

FIGURE 3.5 Large Pumping Station Configuration

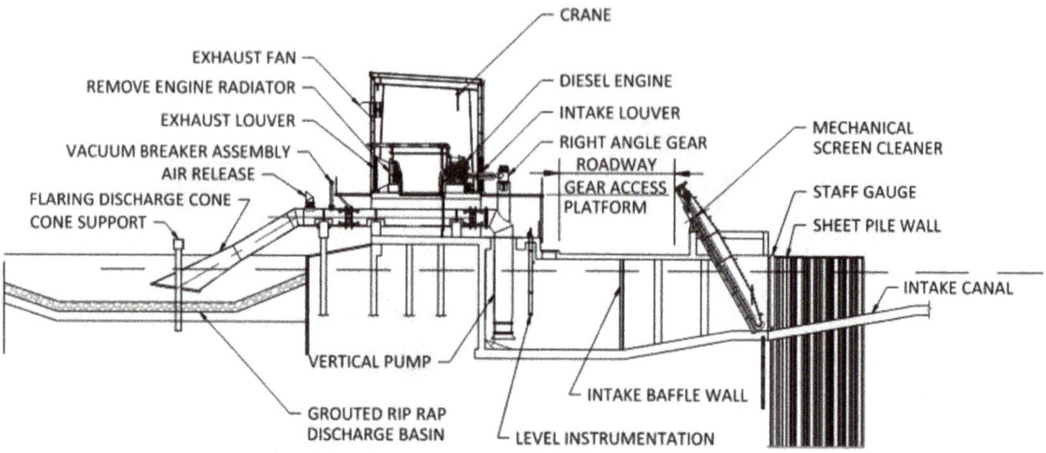

FIGURE 3.6 Section Through Rectangular Wet-Pit-Type Stormwater Pumping Station

emergency functions must be available at all times. Hydraulic conditions will have primary importance in site selection, but site appearance and sound attenuation should also be assessed. Foundation investigations are essential (although they rarely reveal conditions unstable enough to merit rejection of a site); the effects of soil erosion and contaminated runoff on the pumping station influent and the rate of discharge and water quality of the effluent should be evaluated. Enough space must be provided in the area outside the station to accommodate parking and movements of large machinery.

A dependable energy source is essential. The primary source of electrical power for most pumping stations is a public utility. Underground service is preferred for safety and aesthetic reasons, and overhead lines into the station should be avoided because they present potential safety hazards during large equipment operation. For remote sites where electric power may not be available or underground utilities are impractical, consideration should be given for an alternative power source such as a diesel engine.

Some features to evaluate when designing a pumping station depend on the size of the facility, whereas others relate to site conditions, economic considerations, and operating needs. Some of these features to consider are

- building for electrical equipment,
- odor control equipment of chemical storage,
- water for pump washdown (potable or nonpotable),
- crane or trolly hoist for pump removal,
- fencing for privacy and security,

- access driveway for maintenance,
- vent piping for wet well,
- access hatches over the wet well with safety grating or nets,
- chains or cables for pump removal, and
- influent manhole for future bypass pumping access.

Stormwater storage can reduce the required peak capacity of a pumping station. Selection of a peak pumping capacity is a trial-and-error process that considers the inflow hydrograph, available storage, and pump discharge rates.

Pumps are commonly used in storm drainage systems, though the designer is cautioned that they should be used only when gravity flow is not possible.

Stormwater pumping typically uses large pumps that operate at low heads. The main types of pumps include

- vertical (axial and mixed flow),
- submersible (non-clog and column type),
- centrifugal (horizontal non-clog), and
- horizontal (axial flow screw type).

Any of these pumps, besides submersible pumps, can be powered by a motor or engine and can be designed for siphonic flow to reduce head.

2.2 Vertical Pumps

Vertical pumps, the type used most frequently in stormwater pumping stations, are available in many capacities and discharge heads. Single-stage propeller pumps are used for low heads; mixed-flow pumps are used for higher heads. Two-stage propeller pumps are available that will double the head capacity of a single-stage pump.

2.3 Submersible Pumps

Until recently, submersible pumps, except at small stations, have been used as accessory pumps to handle low (dry weather) flows, groundwater infiltration, and cleanup and pump out of the pump pit after a storm. This has been changing as larger submersible pumps have become available that are designed similar to a vertical axial flow pump with the motor coupled to the impeller and inserted into a column pipe. Large stormwater pumping stations can now be designed that rely solely on submersible pumps.

2.4 Centrifugal Pumps

A special type of centrifugal pump is used in dry-pit stations. Designed to handle suspended solids and debris that pass through the screens, these pumps are the end suction, non-clog type. Enclosed impellers and a handhole on the casing permit inspection and removal of foreign material. Small centrifugal pumps combined with upstream storage are well suited to small catchments.

2.5 Horizontal Pumps

Horizontal screw-type pumps can be used in low head of 3 m (10 ft), but at high flowrates of 284 000 L/s (450 000 gpm). These pumps use vacuum pumps and submerged suction and discharge bells to create a siphon in the pump. Once the siphon is reached, the pump can move large amounts of water with a smaller motor than would be required of a pump that either self-primes or pumps to the atmosphere. Given the low lift of these pumps, they must be placed in a low-lying area and can be prone to flooding if the station is not flood-proofed or elevated.

For all pumping stations, the use of pumps with non-clog-type impellers is recommended. Screens or comminutors should be used where the size of solids in the flow stream exceeds the capability of the pumps to handle them. Screen cleaning mechanisms should be used where a large volume of debris is expected, such as the end of an open channel.

It is a good practice for the design engineer to discuss innovative ideas and the unique requirements of each pumping station with pump manufacturers and O&M personnel. Hydraulic Institute standards and publications provided by many of the wastewater and stormwater pump manufacturers have guidelines for dry well and wet well design for wet-pit pump installations. Simplicity and reliability should guide the decision-making process.

3.0 WET WELL CONSIDERATIONS

Wet wells store wastewater and stormwater (typically at smaller stations) before pumping and provide submergence over the pump suction and a smooth transition of flow from the sewer to the pump suction. The wet well influent channel may also house appurtenant equipment (such as comminutors or bar racks) that protect the pump.

Design flowrate determination affects both wastewater and stormwater pumping station configurations, including the wet well. The selected design life of the station is also a key factor in the station configuration. Range of minimum and maximum daily flows and peak hourly flows should be evaluated carefully.

The design of wet wells is important to establish adequate hydraulic flow patterns toward pump suctions. Suction flow requirements for dry-pit pumps are less critical than for wet-pit, axial-flow pumps because the suction pipes to the pump can reduce pre-rotation and other potential problems.

The suction of dry-well pumps is typically a flare fitting on a suction elbow in the wet well or on the vertical wall of the wet well. Concerns for these intakes include approaching velocity and sufficient submergence to prevent a vortex of air into the suction bell. Suction piping should be designed for the maximum pumping capacity of each pump.

The wet well configuration for wet-pit, axial-flow pumps is critical to ensure even flow to the suction bell of the pump without surface or submerged vortices. Items to be considered include currents created in the wet well by the incoming flow, flow past an inoperative pump to an operating pump, and submergence over the suction bell. In many installations, baffles to interrupt high-velocity flow and training walls to direct flow to a pump suction are required. For large or complex pumping stations, it is advisable to study the hydraulic flow patterns in a reduced-scale hydraulic model of the wet well. Sometimes, model studies for similar designs can be used.

It is important to reduce or eliminate free fall of water into the wet well to prevent entrained air from entering a pump suction. Entrained air forms a large bubble at the eye of the impeller that can cause vibration, reduce pump capacity, and even stop the pump from functioning by air binding it.

To allow air bubbles to rise to the surface, there must be sufficient distance from the point of free fall to the pump suction inlet. Inlet drop pipes or baffling help divert air from the pump suction. Also, a steeply sloping wet well floor helps separate the air from inflowing liquid.

In large systems, especially those with inlet screens, the high wet well level should be set at the highest possible level without surcharging the influent sewer. This reduces free fall at low wet well levels.

The required storage volume in the wet well depends on the method of pumping used. Pumping stations with on–off control of the pumps require sufficient storage in the wet well to prevent the pump motors from short cycling and overheating. For dry-pit motors, a maximum of six starts per hour is recommended for motors 20 kW (25 hp) and smaller, four starts per hour for motors between 25 and 75 kW (33 and 100 hp), and two starts per hour for motors between 100 and 200 kW (135 and 270 hp). Larger, costlier motors that allow more starts per hour are available, but these motors are slightly less efficient than conventional motors. The motor manufacturer should be consulted on the allowable starting frequency of such motors.

Submersible pumps that have motors designed to operate in liquid are typically designed for up to 10 starts per hour for motors less than 40 kW (50 hp), six starts per hour for motors 40 to 80 kW (50 to 100 hp), and four starts per hour for motors 80 kW (100 hp) and up. Some larger motors are cooled by circulating some of the pumped fluid through a cooling jacket. The pump manufacturer should be consulted for allowable frequency of starts, especially if the motor is not submerged.

Elapsed time between starts is a function of the pump discharge rate and the inflow to the wet well. In the case of a multispeed pump or multiple pumps, it is determined by differences among step flowrates when the inflow is between the two steps. The time between starts is at a minimum when the inflow is 50% of the pump rate halfway between two steps. The general formula for computing the required wet well storage is

$$V = tQ/4 \tag{3.1}$$

Where

V = storage volume, m^3 (gal);
t = required time between starts, minutes; and
Q = pump discharge capacity, or the difference in flowrate between steps, m^3/min (gpm).

Many public agencies specify a maximum storage time in a wet well to reduce the potential for odors and septic wastewater. A maximum storage retention time of 30 minutes at average flow is a common requirement. Shorter retention time can conflict with requirements to prevent short cycling. In such cases, multispeed pumps (including variable-frequency drives) or multiple pumps should be considered.

Variable-speed pumps require little storage time because they match the incoming flowrate, thus eliminating frequent starts and stops. Most controls are set to maintain maximum speed at high wet well levels and minimum speed at low wet well levels. The vertical distance between the two levels and the resulting storage in the wet well must be sufficient to allow stable operation of the speed controls in the event of sudden changes in incoming flow.

Wet wells should be configured to prevent accumulation of solids settling that could occur independent of the pump cycling time. For large pumping stations, compartmentalization for the purpose of guiding flow to each pump suction is an option.

Sewer gas generated in the sewer and volatile material spilled into the sewer are transported and exposed to the atmosphere in the wet well. Therefore, all equipment (particularly electrical motors) in the wet well must be of nonvolatile, explosion-proof construction.

If the station has bar racks or screens, another consideration is the control of flow velocity through the bars. Velocity through the bars is a function of flowrate and wetted area and can be controlled by the wet well levels.

The velocity through the clear opening between bars should be between a maximum of 1.2 m/s (4 ft/sec) and a minimum of 0.4 m/s (1.2 ft/sec). At high wet well levels (high speed), both the flowrate and the wetted area are at a maximum. At low wet well levels (minimum speed), the flowrate is at a minimum, but the wetted area is also at a minimum.

4.0 AESTHETIC AND SAFETY CONSIDERATIONS

Aesthetic considerations can provide pleasant surroundings for the public and a comfortable atmosphere for station O&M personnel. Preliminary stages of pumping station design should evaluate numerous structural, architectural, practical, and safety constraints, including the following:

- Building material and configuration compatible with neighboring developments
- Adequate ventilation for workers
- Control of ventilation fans could be building temperature, carbon dioxide, or occupancy sensors
- Continuous or timer-controlled air changes, which may be required to meet local and national safety regulations
- Separate ventilation for dry and wet wells
- Provisions for emergency eyewash and chemical leak alarm linkage with local fire station (if chemicals such as hydrogen peroxide and chlorine are used for odor control or other dangerous liquids are stored on-site, including battery acid and diesel fuel)
- Building construction of vandal-proof material
- Gas migration barrier between wet well and dry well (interior wet well coating [to guard against corrosion] typically offers this protection)
- Effective odor mitigation system to avoid offending neighbors
- Unobstructed sight paths for neighboring properties
- Potential for flooding from backflows because of high water level at the station wet well
- Site drainage and erosion control integrated with good landscaping design
- Choice of fencing to satisfy local zoning and historical regulations
- Use of parapet walls to hide certain equipment (such as heating, ventilation, and air-conditioning [HVAC] ducting and generator muffler exhaust piping)

- Sprinkler system, which may be required to meet local codes or if fuel is stored on-site
- Need for an on-site restroom
- Sloped dry well floors to avoid accumulation of water
- Noise control through use of acoustical materials at emergency generators and other sensitive locations
- Landscaping with native plants to blend the station into the neighborhood
- Exterior lighting consistent with that of neighbors
- Adequate parking and easy access for delivery of chemicals, maintenance of equipment, and service or removal of pumps, motors, drives, and auxiliary equipment
- Safety grating or nets at all access hatches
- Station accessibility during a storm event

5.0 ARCHITECTURAL DESIGN CONSIDERATIONS

Station arrangements, building layouts, and safety features must satisfy the requirements of local, state, and national codes (such as those of the Occupational Safety and Health Administration). The above-grade structure is designed to provide access to the lower levels of the pumping facility containing the electrical controls and ancillary equipment; the station design itself must be compatible with local zoning and site-approval criteria. Factors include plans for lighting (discussed in more detail in Chapter 6), plumbing, and HVAC systems.

In residential and commercial areas, there should be adequate distance between buildings, if possible. Concrete masonry blocks (plain face or split face) and bricks are suitable building materials, as is stucco construction (wood or steel stud framing). Windows (with steel wire-reinforced glazing for protection against vandalism) and false windows can be integrated with the building. Using the right approach, each pumping station can be pleasing and acceptable to the neighborhood property owners.

Adequate lighting with convenient switch locations should be provided for all stations. Skylights to enhance natural lighting are a possibility. Shadows and glare should be avoided near machinery and at floor openings. Deep dry wells should have multilevel light fixtures located such that maintenance will not be a safety concern. A motion-actuated exterior lighting system is an option.

To ensure adequate building drainage, designs should include a minimum lower floor slope of 10 mm/m (1/8 in./ft), and floor gutters are recommended for areas of chronic wetness. Floor drains should be provided in restrooms and adjacent to pumps and generator oil drain outlets (to collect accidental spills during maintenance). An exterior oil–water separator sump should be considered, and sump pumps actuated by level switches should be used to transfer the drainage from floor sump to the wet well. Check valves should be installed in each sump pump discharge line to avoid the backflow of wastewater to the sump. The main sump pump discharge piping should connect to the wet well at a point higher than the high-water alarm level.

The station site should be graded, landscaped, and designed with properly vented drainage to conform to the surrounding properties and keep them free of drainage overflow.

For sensitive electronic control equipment, localized HVAC cooling systems can be used to mitigate potential damage from heat generated by large motors and emergency generators. Proper separation of HVAC ducting and electrical panels helps prevent interference; and, to screen rooftop HVAC equipment, exterior treatment such as using a parapet is an option.

6.0 STRUCTURAL DESIGN CONSIDERATIONS

Station hydraulic structures (substructures) can either be custom designed or pre-engineered (prefabricated package pumping stations). For small stations, a precast or cast-in-place concrete manhole (a minimum of 1.75 m [5 ft] in diameter) could serve both as wet well and submerged pumps housing.

The depth of incoming sewers or storm drainage channels, and sometimes foundation conditions, will determine the depth of the station substructure below the ground level. Surface conditions, environment, and necessary elevation to avoid flooding should determine the level of the operating floor, the type of superstructure, and the required exterior finish and protection.

The following criteria must be considered in designing the lower structures:

- All hydraulic structures must be designed to withstand uplift forces from high groundwater conditions and earthquakes. The designer will need to consider uplift from groundwater during construction or maintenance when the wet well will be empty.
- For stations more than 9-m (30-ft) deep, an intermediate (mezzanine) floor can be added to increase structural rigidity and provide support for drive shaft bearings (if the motor and drive unit is placed at the

upper floor) and lighting fixtures. This extra floor space could also be used for storage. The same approach can be taken for a station with an operating floor that has been elevated above grade to meet local floodplain requirements, and adequate clearance should be provided on the ground floor for storage of sacrificial materials.

- For excavation, backfill, sump dewatering, and foundation requirements, reference the soil profile investigation. For deep station construction and in areas of high groundwater, bidding documents should not be completed until sump pump and well point dewatering operations are approved by geologists, hydrogeologists, geotechnical engineers, and dewatering subcontractors. The engineer and client should determine if the contract is to provide specific excavation support system details or if the documents are intended to include only design criteria related to the development of an excavation support structure. To secure the lowest bid price, the client should allow specific design of the excavation support system to be the responsibility of the contractor.
- All entrances and non-watertight openings in the station should be at elevations above the flood protection elevation.
- Station substructures typically should be made of reinforced concrete. For small, prefabricated stations, steel plate construction and fiberglass are acceptable. American Concrete Institute and American Institute of Steel Construction standards should be followed, in addition to the Uniform Building Code and applicable local codes.
- Designs must provide for practical ways to remove equipment from the pumping station by including access openings and hatches, whether it be a separate access hatch per pump or a single opening. Skylights can serve as access hatches in large, deep, vertical pumping stations.
- For small stations, A-frames and dollies are needed to lift pumps, motors, and gears. For large stations, a monorail or a traveling bridge crane can be used to lift heavy items, though such hoisting equipment needs adequate space in which to operate.
- All entrances should have handrails and gratings, with nonslip treads standard for all stairways.
- Access hatches and safety ladders or stairs should allow for proper inspection and maintenance of both dry and wet wells. Ladders must be provided with safety cage or harness arrangements, with stainless steel or aluminum ladders recommended for the wet well. Access hatches should be provided with safety grates or nets.
- Because wastewater pumping stations continuously are exposed to corrosive environments, wet wells can be equipped with a chemical

feed system for odor control. Wet well walls and the bottoms of top slabs should be coated or lined with an epoxy or polymer. Wet wells must be provided with positive-forced ventilation, and their floors typically are sloped to guide the solids toward the pump suction elbows.

- Adequate ventilation is essential to protect the dry well from moisture, especially in the presence of high groundwater.
- Consideration should be given to the use of package stations, which are factory-fabricated enclosures with special valves and fittings designed for compactness. The enclosure typically is constructed of steel, with various options of coating and cathodic corrosion protection systems. Prefabricated reinforced concrete stations are also available, and some manufacturers offer fiberglass enclosures. Package stations have the advantages of low initial cost and rapid installation and can be relocated. However, the cost of rehabilitating and retrofitting piping inside pumping stations should be evaluated carefully. Disadvantages of package stations include a limited operating life because of corrosion, restricted working space, and inflexible piping for pump replacement and station expansion.
- Structural design requirements for built-in-place concrete stations should also apply to package stations, which require a certain amount of engineering design analysis, such as design of a separate wet well and a foundation slab. Specific tie-down designs could be required for lightweight packaged stations at high groundwater locations.
- Access to the inside of a wet well for removal of sand, grit, grease, and large objects should be considered. The addition of bar screens, comminutors, and other pretreatment devices affects wet well design.

7.0 REFERENCE

U.S. Army Corps of Engineers. (2020). *Mechanical and electrical design of pumping stations* (EM 1110-2-3105).

8.0 SUGGESTED READINGS

Hydraulic Institute. (2018). ANSI/HI standard for rotodynamic pumps for pump intake design 9.8.

Jones, G. M., & Sanks, R. L. (2008). *Pumping station design* (3rd ed.). Butterworth-Heinemann.

4

Pumping Systems

Pumps and pumping systems play a vital role in wastewater and stormwater systems. Therefore, it is imperative that design engineers produce well-engineered pumping stations that operate reliably and efficiently.

This chapter provides design criteria for wastewater and stormwater pumping stations that are located in collection systems. These vary from small, simple stations in remote areas of collection systems to large, complex pumping stations that discharge all or a significant portion of a collection system's wastewater to a water resource recovery facility (WRRF).

Also discussed in this chapter are characteristics of pumps and drives, fundamentals of system hydraulics, the use of system head-capacity curves, and the selection of equipment for reliable, efficient pumping systems to meet the needs of the system head-capacity curve.

Pumping station design must recognize and blend many requirements into an integrated scheme. Some of these requirements are discussed further in other chapters of this manual.

1.0 WASTEWATER AND STORMWATER PUMPING STATIONS

There are similarities and differences between wastewater and stormwater pumping stations. Both receive flow from a drainage system and discharge the flow. An obvious difference is that wastewater pumping stations are in continuous operation, whereas a stormwater pumping station may be idle for considerable periods of time between storm events.

A key factor in the design of a pumping station is the type of sewer system that delivers flow to the station. The three types of sewer systems are sanitary, storm, and combined sewers. A sanitary sewer carries inflow and infiltration, but does not carry stormwater. Sanitary wastewater typically does not contain large solids (more than 75 mm [3 in.] in diameter) or large quantities of grit, but can contain rags and stringy material that clog pumps.

Storm sewers collect surface runoff, groundwater infiltration, and inflow. Stormwater flow typically contains large solids, stringy material, and large quantities of grit.

Combined sewers carry sanitary wastewater during dry weather periods, but carry both sanitary wastewater and stormwater during wet weather periods.

Both wastewater and stormwater pumps often require screens or other protective devices to remove solids or stringy material that might clog pumps. Wastewater pumps, especially those discharging to a WRRF, often remove the maximum quantity of material and reduce the load on the treatment process by having screens with only small openings between bars. Stormwater pumping stations, however, typically have large, clear openings between bars that protect the pump yet reduce the quantity of screenings that must be handled at the station. Chapter 8 presents additional information on screens and comminutors.

The end suction centrifugal pump is the most frequently used type in wastewater pumping stations. Vertical pumps are the common type for stormwater applications. However, stormwater pumping stations generally have higher capacities at lower heads than wastewater pumping stations and often are only lift stations that receive flow from an interceptor and discharge to an open channel. Therefore, Archimedes screw pumps and vertical mixed- or axial-flow wet-pit pumps are common in stormwater pumping stations.

Electric motor-driven pumps are more common than engine-driven pumps in both wastewater and stormwater pumping stations. Stormwater stations frequently are idle for extended periods of time but, when operating, have high power draw. This duty cycle may favor the use of engine-driven pumps. In selecting the type of engine drive, the higher operations and maintenance costs of the drive should be compared to the demand and energy costs for electrical power. Social and environmental effects should also be considered such as noise, air pollution, risk of fuel spill, ambient temperature effect on fuels, and so on.

Figure 4.1 presents a depiction of a typical medium-sized pumping station. The station houses four pumps driven by shafting, with the motors mounted on the floor above the pumps.

FIGURE 4.1 Medium-Sized Wastewater Pumping Station

2.0 PUMPING SYSTEM HYDRAULICS AND DEFINITIONS

Pumps are used to transport water from one point to another. To accomplish this, energy must be expended. The amount of energy expended depends on the quantity (flowrate), resistance to flow (the head the pump must overcome), and efficiency of the pump and drives. The terms associated with pumping systems include *capacity*, *head*, *pump efficiency*, and *power requirements*.

2.1 Capacity

Capacity is the volume of liquid pumped in a unit of time. The capacity typically is expressed as liters per second (L/s) or cubic meters per second (m^3/s) (gallons per minute [gpm] or million gallons per day [mgd]).

2.2 Head

A pressure term, *head* is the elevation of a liquid surface above or below a reference datum elevation. For instance, if an open-ended tube were extended vertically from a pipe under pressure, the head would be the vertical distance from the centerline of the pipe to the free water surface in the tube. Head typically is expressed in terms of meters of fluid (feet of fluid).

Head can refer to both pump head and system head. Pump head is the height the pump can raise the liquid being pumped at a given flowrate. System head is the height to which the liquid must be raised to overcome the resistance of a pipe system at a given flowrate.

Many types of head are included in the analysis of a pumping system, as illustrated in Figure 4.2. The datum elevation typically is taken as the centerline of the pump. Heads above datum are taken as positive heads and heads below datum are negative.

Static head (H_{stat}) is the difference in elevation between the free liquid surface on the discharge side of the pump and the free liquid surface on the suction side of the pump. Therefore, H_{stat} is the algebraic difference of the discharge static head (h_{sd}) and the suction static head (h_{ss}). Suction and discharge static heads are the difference in elevation between the suction, or discharge liquid surface, and the centerline of the pump.

Velocity head is the kinetic energy contained in a fluid in motion and is given by the following formula:

$$H_v = \frac{v^2}{2g} \tag{4.1}$$

Where

H_v = velocity head, m (ft);
V = velocity of fluid, m/s (ft/sec); and
g = acceleration because of gravity, 9.81 m/s^2 (32.2 ft/sec^2).

The *energy grade line* refers to a line that represents the energy head of the system in meters or feet and is the sum of the pressure head, velocity head, and elevation head. The *hydraulic grade line* refers to a line that represents the hydraulic head of the system in meters or feet and is the sum of the pressure head and the elevation head of the system.

Friction head (H_f) in a piping system is the head of liquid required to overcome the resistance to flow caused by the drag of the pipe wall on the

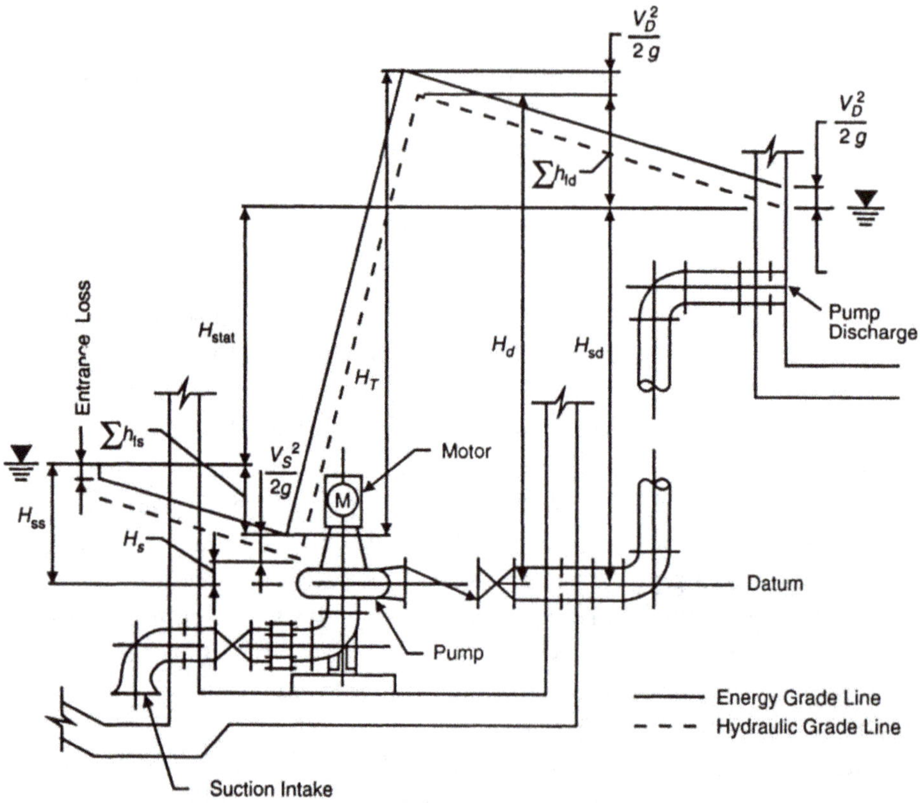

FIGURE 4.2 Definition Sketch for Pump Heads

Note. $V_S^2/2g$ = suction velocity head, $V_D^2/2g$ = discharge velocity head, H_{ss} = suction static head, H_{sd} = discharge static head, Sh_{fs} = total suction friction losses, Sh_{fd} = total discharge friction losses, H_s = suction head, H_d = discharge head, H_{stat} = net static head, and H_T = total pump head

fluid. As with static head, the friction head can be separated into suction friction losses (h_{fs}) and discharge friction losses (h_{fd}). Friction head is also known as a *major head loss*.

Minor head losses (h_m) are losses required to overcome the resistance to flow through fittings and valves. Minor losses can be added to friction losses because both are caused by the resistance to flow in a pipe system. Figure 4.2 shows friction and minor losses in the suction combined, shown as Σh_{fs} and, in the discharge, as Σh_{fd}.

Total head is the algebraic total of the suction and discharge heads as given by the following equation:

$$H_T = H_d - H_s \qquad (4.2)$$

and

$$H_d = H_{sd} + \Sigma h_{fd} + H_{vd} \tag{4.3}$$

$$H_s = H_{ss} - \Sigma h_{fs} \tag{4.4}$$

Where

H_T = total dynamic head (TDH), m (ft);
$H_d(H_s)$ = total discharge (suction) head measured at the pump nozzle with static head based on pump centerline, m (ft);
$H_{sd}(H_{ss})$ = discharge (suction) static head, m (ft);
$\Sigma h_{fd}(\Sigma h_{fs})$ = the total of the discharge (suction) friction and minor losses, m (ft); and
H_{vd} = discharge velocity head at the end of the force main, m (ft).

If the suction and discharge static heads are combined, Equation 4.3 can be rearranged as

$$H_T = H_{stat} + \Sigma h_{fd} + \Sigma h_{fs} + H_{vd} \tag{4.5}$$

Where

H_{stat} = total static head, m (ft).

2.3 Pump Efficiency and Power Requirements

The amount of energy required for pump operation is proportional to the rate of flow, the head required to overcome system resistance, and the efficiency of the pump. This energy, or power, is measured in kilowatts (kW) or horsepower (hp).

In the metric system, when pumping water, the pump input power is

$$P_s = \frac{QH_T}{0.102E_p} \tag{4.6}$$

Where

P_s = input shaft power, kW;
Q = flowrate, m^3/s;
H_T = total head, m; and
E_p = pump efficiency, dimensionless.

In English units, when pumping water and the flow is given in gallons per minute (gpm), the pump input brake horsepower is

$$P_s = \frac{QH_T}{3960E_p} \tag{4.7}$$

Where

P_s = input shaft power, hp;
Q = flowrate, gpm;
H_T = total head, ft; and
E_p = pump efficiency, dimensionless.

It should be noted that Equation 4.7 is input energy to the pump. To compute the total power of the pump and drive, the pump input power must be divided by the efficiency of the driver.

3.0 PUMPS

Pumps typically are categorized as kinetic energy pumps or positive displacement pumps. Kinetic energy pumps produce high-velocity kinetic energy in the pump that is converted to pressure. Positive displacement pumps produce pressure by applying force directly to the fluid.

Kinetic energy pumps include centrifugal, peripheral, and special pumps. Centrifugal pumps are the most common type of wastewater pumps and can discharge large volumes of water and pass the solids normally found in wastewater or stormwater. Peripheral turbine-type pumps and special pumps (such as hydraulic ram or jet pumps) are not used in wastewater service because they require close clearances and cannot pass wastewater solids.

Positive displacement pumps include reciprocating piston and plunger, rotary (progressive cavity, gear and vane, peristaltic, and screw pumps), and blow case pumps. Many positive displacement pumps are used for pumping sludge and scum at WRRFs but are not common in wastewater and stormwater pumping stations because of their limited capacities and inability to pass large solids.

The two types of positive displacement pumps that are used in wastewater and stormwater pumping stations are the Archimedes screw pump and the pneumatic ejector. The Archimedes screw pump is a lift pump that can discharge large quantities of flow and the pneumatic ejector is used to discharge small flows that contain wastewater solids.

4.0 CENTRIFUGAL PUMPS

The centrifugal pump is the most common type found in wastewater pumping stations. Centrifugal pumps consist of two main parts: a rotating impeller and the stationary pump casing. The impeller is driven by the pump shaft and transfers energy to the fluid being pumped. This energy produces flow and head. The casing directs the flow to the pump discharge and converts some of the flow velocity to head.

4.1 Pump Classification

Centrifugal pumps are classified as radial-flow, mixed-flow, and axial-flow pumps, where the type of pump describes the shape of the impeller and the discharge flow pattern in relation to the pump shaft. Flow enters the impeller parallel to the pump shaft. Discharge from a radial-flow impeller is perpendicular to the pump shaft, discharge from an axial-flow pump is parallel to the pump shaft, and discharge from a mixed-flow impeller is at an angle to the shaft.

The shape of the casing will also depend on the type of pump. Radial and some mixed-flow pumps have volute-type casings that collect the flow as it leaves the impeller and guides it to the discharge nozzle, which is perpendicular to the pump shaft. Axial-flow pumps and some mixed-flow pumps are fitted with diffusers that straighten the flow as it leaves the impeller so that the flow is parallel to the shaft.

The shape of the impeller determines the flow pattern. A type number, known as *specific speed*, has been developed that characterizes pump type and performance. Specific speed is based on pump performance at its best efficiency point (BEP), and pumps of the same specific speed will have similar head-capacity curves. The BEP is the point along the pump performance curve where efficiency is the highest and is a unique characteristic to each pump. The formula for specific speed is

$$N_s = \frac{NQ^{0.5}}{HT^{0.75}} \tag{4.8}$$

Where

N_s = specific speed;
N = pump rotative speed, r/min (rpm);
Q = flowrate at BEP, m^3/s (gpm); and
H_T = total head at BEP, m (ft).

The specific speed in metric units can be multiplied by 51.6 to be converted to English units. When calculating the specific speed for double suction pumps, only the flow through one side of the impeller, or 50% of the total flow, should be used. However, double-suction pumps are not typically used in wastewater or stormwater service.

Specific speed has value because it indicates the type of pump and is useful in predicting the performance of a pump. The Hydraulic Institute

Standards have tables and curves that predict, based on pump specific speed, design and cavitation characteristics and abnormal operation of pumps.

Radial-flow pumps have specific speeds that range from 10 to 80 (500 to 4000 in English units), mixed-flow pumps have specific speeds from 60 to 175 (3000 to 9000 in English units), and axial-flow pumps have specific speeds of more than 175 (9000 in English units).

Figures 4.3 and 4.4 show the approximate shape of the impeller and the expected efficiencies for various-capacity pumps as the pump specific speed varies. Figure 4.3 is in metric units and Figure 4.4 is in English units. Mixed-flow impellers in the specific speed range of 60 to 125 (3000 to 6500 in English units) are available with volute-type casings with discharge perpendicular to the shaft and are known as *mixed-flow volute pumps*. Mixed-flow impellers in the specific speed range of 75 to 175 (4000 to 9000 in English units) are available with diffuser-type casings for discharge parallel with the pump shaft and are known as *mixed-flow propeller pumps*.

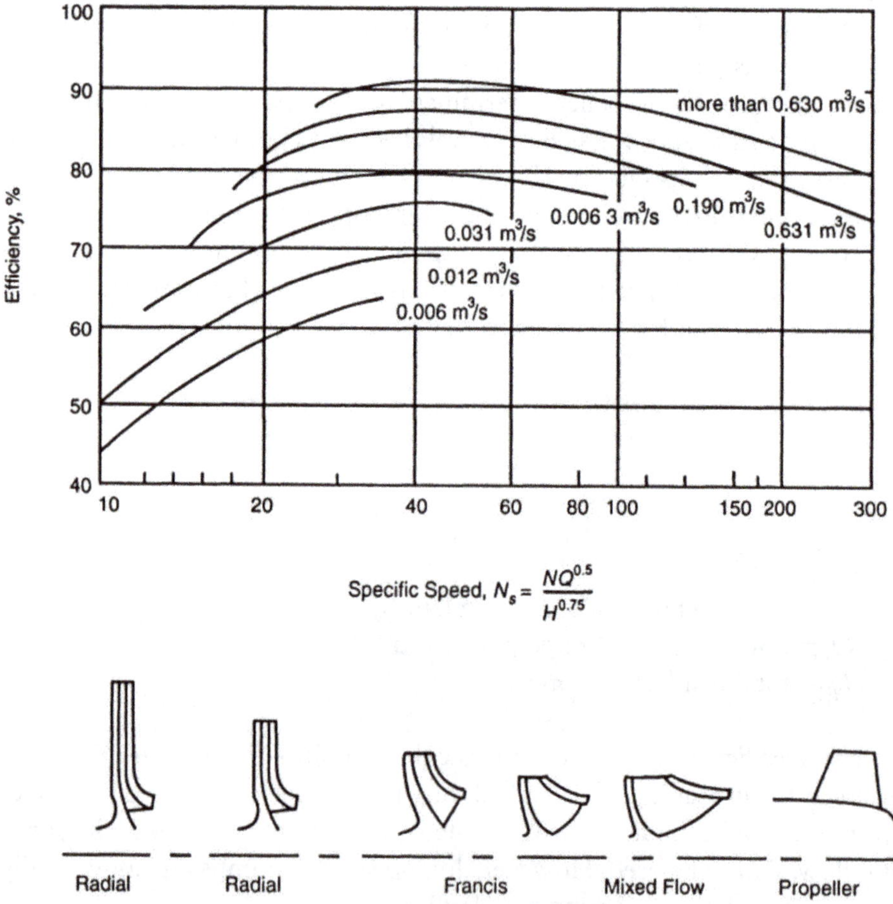

FIGURE 4.3 Specific Speed, Impeller Shape, and Efficiency—Metric Units

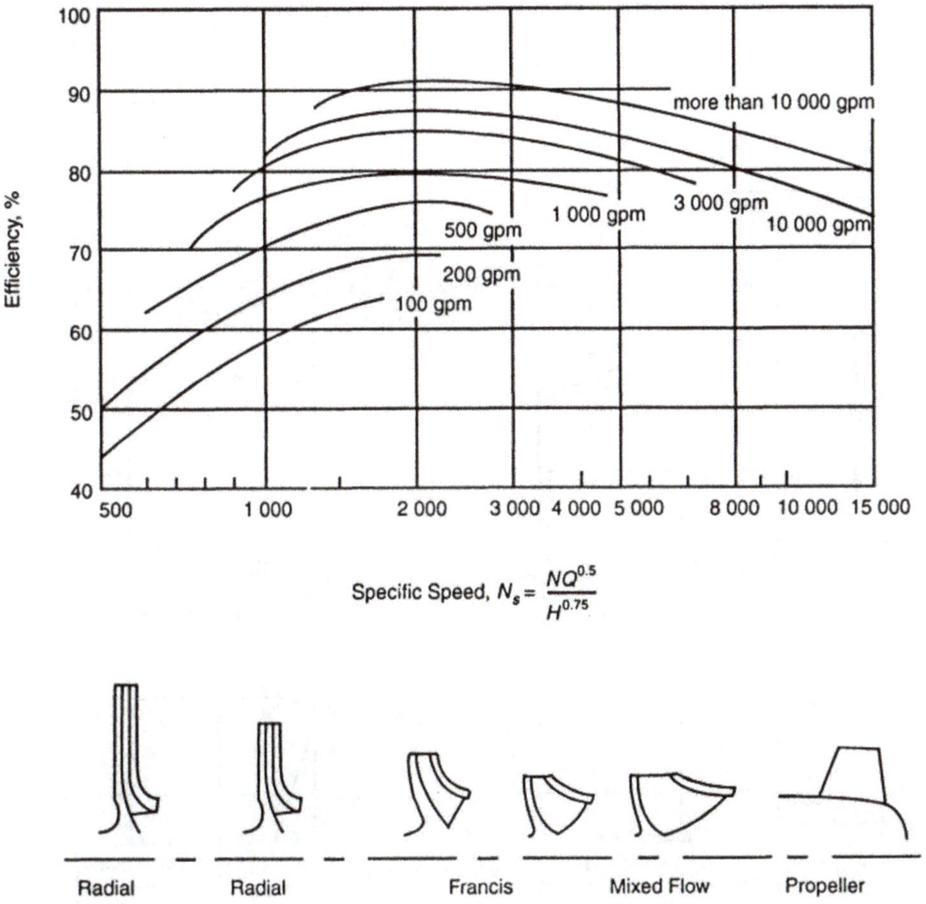

FIGURE 4.4 Specific Speed, Impeller Shape, and Efficiency—English Units

Figure 4.5 illustrates typical head capacity, power input, and efficiency curves for various specific-speed pumps. As pump-specific speed increases, the pump horsepower at shutoff (zero flow) varies from non-overloading (less than the pump horsepower at the BEP) for radial-flow pumps to extreme overloading (more than the pump horsepower at the BEP) for axial-flow pumps.

End-suction radial pumps and mixed-flow volute pumps are most commonly used in wastewater and stormwater service. Double-suction pumps are not typically used because rags can clog the pump by wrapping around the pump shaft at the eye of the impeller.

Axial-flow propeller and mixed-flow, wet-pit pumps are capable of discharging large quantities of flow at low heads and typically are less expensive than volute pumps. These pumps are common in stormwater systems but are not typically used in wastewater systems unless well protected by

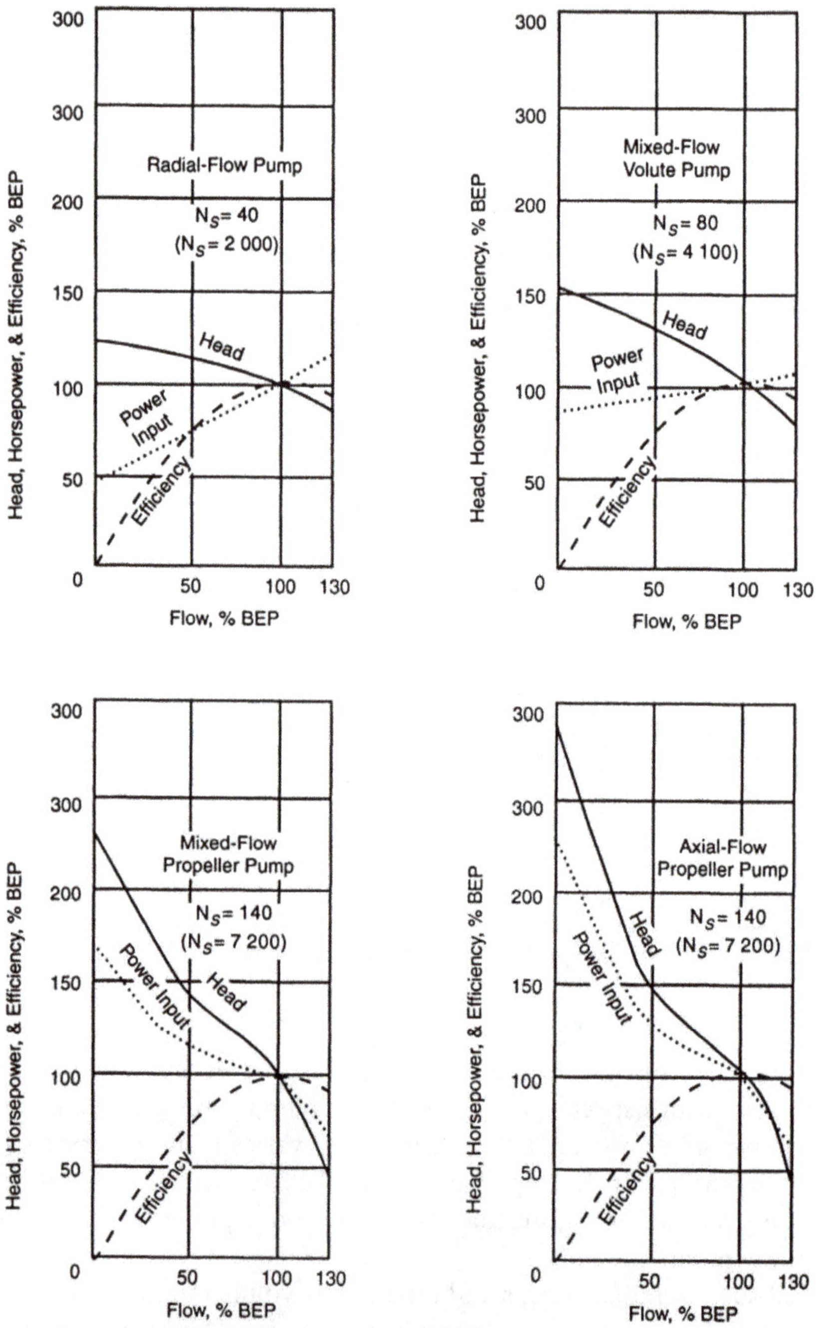

FIGURE 4.5 Centrifugal Pump Characteristic Curves for Various Specific Speeds (BEP = best efficiency point) (hp × 745.7 = W)

screens because rags and stringy material can clog the pump by building up on the inlet or diffuser vanes.

4.2 Impellers

The pump impeller is driven by the pump shaft and transfers energy to the fluid being pumped. Wastewater and stormwater pumps must be able to pass solids that enter the system. Domestic water closets can pass a 65-mm (2.5-in.) spherical solid. Therefore, many codes require wastewater pumps to pass a 75-mm (3-in.) solid. Also, rags and other stringy material enter wastewater, often presenting more of a clogging problem than solids.

Because of the increased presence of rags and disposable wipes in domestic wastewater, modern impeller designs have trended toward including a cutting edge. Some of these impellers cannot "pass" a spherical solid, as indicated above, but are able to "manage" the solid by breaking it up. This should be considered when meeting local governing criteria because a variance or exception may be required based on this definition.

Impellers are available in open, semi-open, and enclosed constructions. A semi-open impeller has one shroud on the back of the impeller and the exposed blades, or vanes, of the impeller fit closely to the stationary pump casing. This close clearance prevents recirculation from the discharge to the suction of the pump. A closed impeller has shrouds on both sides of the impeller, so close clearance is required only at the eye of the impeller between the lower shroud and the pump suction cover. An open impeller has no shrouds on the vanes.

Wastewater and stormwater radial-flow pumps typically have closed impellers to protect the edges from being eroded by grit and other abrasives. Such erosion would open up the clearance between the vane and the casing, reducing pump efficiency.

An exception is the torque-flow or vortex-type pump, upon which the impeller is semi-open and located out of the stream of flow passing through the pump. The rotation of the impeller produces a vortex flow in the pump casing that induces flow through the pump. Because solids and stringy material do not pass through the impeller, the pump is unlikely to clog. However, these pumps are less efficient than enclosed impeller pumps.

4.3 Pump Characteristic Curves

Pump characteristic curves are furnished by manufacturers to provide information on the performance of their pumps. Pumps of comparable size and specific speed are produced by many manufacturers, but there are differences in pump performance because of design modifications.

Pump characteristic curves show the pump discharge head, required horsepower, and efficiency plotted against the pump discharge rate. They also indicate the net positive suction head required by the pump ($NPSH_R$) to prevent cavitation. Cavitation and $NPSH_R$ are discussed later in this chapter.

The manufacturer typically provides different diameter impellers in one size casing that cover a range of discharge heads. Therefore, the pump characteristic curve shows a series of parallel curves for a single-size pump.

Note that the $NPSH_R$ by the largest impeller at a given flowrate is different from that by the smallest impeller at the same flowrate, and many manufacturers do not show the effect of impeller diameter on the $NPSH_R$ by the pump.

Manufacturers present characteristics in different ways. Figures 4.6 and 4.7 present examples of typical pump curves. Figure 4.6 shows the $NPSH_R$, brake horsepower (BHP), and efficiency all plotted on the head-capacity curve. Figure 4.7 provides two separate curves for $NPSH_R$, one each for the largest and smallest diameter impellers. Note that the specific speed of the 600-mm (24-in.) impeller in Figure 4.6 is 38.5 (2000), while the specific speed for the 480-mm (19-in.) impeller is 50 (2595).

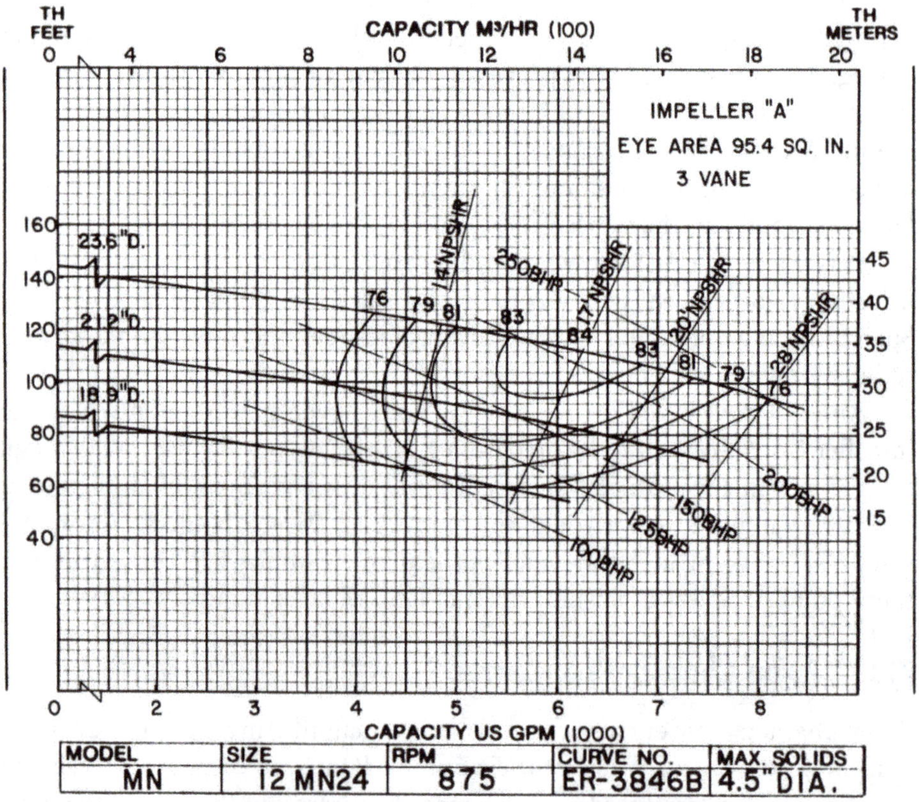

FIGURE 4.6 Centrifugal Pump Curve (Superimposed $NPSH_R$ Curves)

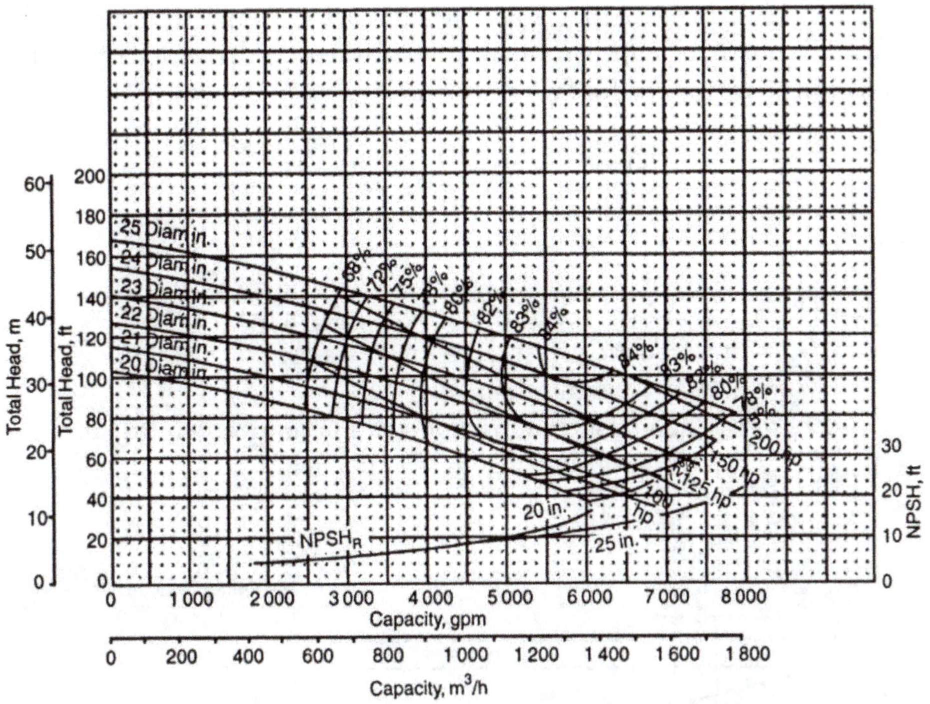

FIGURE 4.7 Centrifugal Pump Curve (Separate $NPSH_R$ Curves)

Note. hp × 745.7 = W, ft × 0.3048 = m, and in. × 25.4 = mm

4.4 Cavitation and Net Positive Suction Head

Cavitation in a pump occurs when the absolute pressure at the eye of the impeller is less than the vapor pressure of the fluid being pumped, causing the fluid to boil and form vapor bubbles. As these bubbles move into a zone of higher pressure, they collapse and the surrounding fluid rushes violently to fill the void, producing a hammering action that can cause vibration and pitting of the pump impeller. Cavitation results in the loss of pump capacity and efficiency and causes physical damage to the pump. When dealing with cavitation, the engineer must be familiar with net positive suction head (NPSH), which is based on absolute pressure. Two different NPSH terms are used. The first is the *net positive suction head available in the system* ($NPSH_A$), which is the absolute liquid pressure produced by the system at the eye of the impeller at a given flow. The second term is the *net positive suction head required by the pump* ($NPSH_R$) to prevent cavitation at a given flow. Cavitation occurs when the NPSH required by the pump at the eye of the impeller is greater than the NPSH available in the system.

The $NPSH_A$ is the total absolute head above the vapor pressure of the fluid at the eye of the impeller. Figure 4.8 illustrates the derivation of $NPSH_A$ and is calculated by the following formula:

$$NPSH_A = H_A + h_{ss} - \Sigma h_{fs} - H_{vp} \tag{4.9}$$

Where

$NPSH_A$ = net positive suction head available, m (ft);
H_A = atmospheric pressure, m (ft) of water;
h_{ss} = suction static head, m (ft);
Σh_{fs} = the sum of suction friction and minor losses, m (ft); and
Hvp = liquid vapor pressure, m (ft) of water.

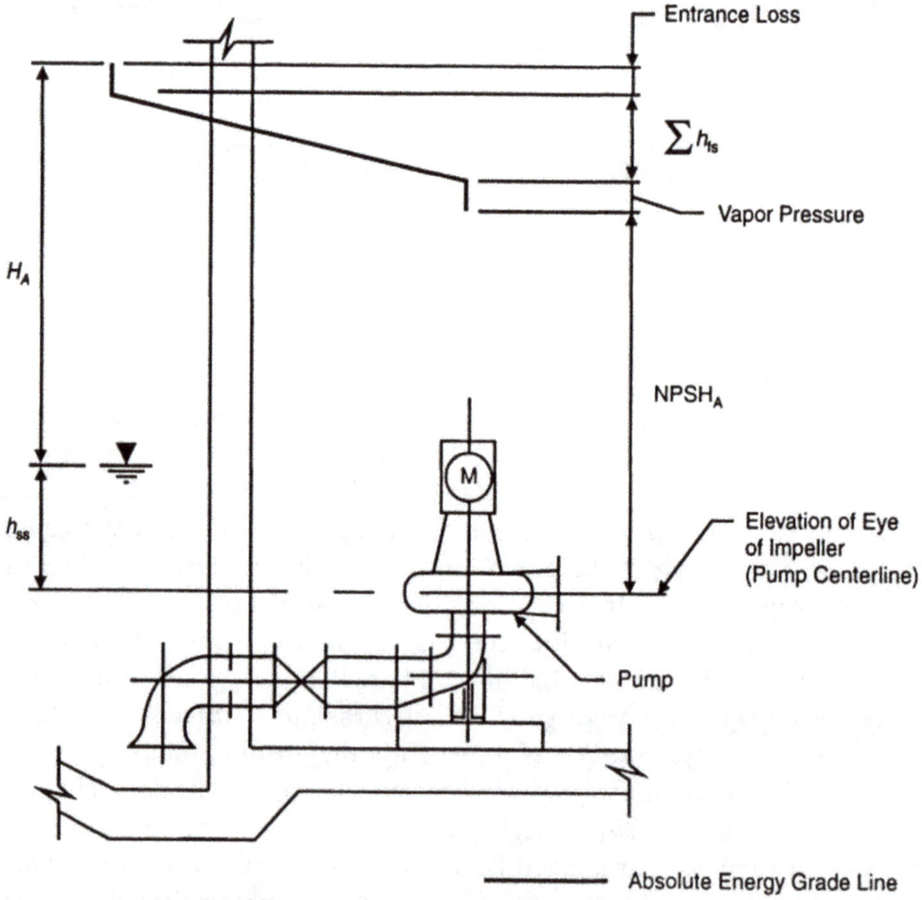

FIGURE 4.8 Definition Sketch for System $NPSH_A$

Note. $NPSH_A$ = net positive suction head available; H_A = atmospheric pressure; h_{ss} = suction static head; and Sh_{fs} = total suction friction losses

Atmospheric pressure varies with elevation (Table 4.1) and liquid vapor pressure varies with the temperature of the pumped fluid (Table 4.2).

The $NPSH_R$ is a function of the impeller design and is provided by the manufacturer; it is typically given on pump curves. The $NPSH_R$ by the largest diameter impeller on the pump curve is different from the $NPSH_R$ by the smallest diameter impeller at a given flow (Figures 4.6 and 4.7).

The requirement to select a pump that will operate with less $NPSH_R$ by the pump than the $NPSH_A$ provided by the system is critical. The $NPSH_R$ should be checked at the extreme of pump operation with minimum static head (high wet well to low discharge water surface), with friction losses computed for new pipe.

The manufacturer's $NPSH_R$ testing is done under ideal conditions, and field installation of a pump may include elbows or valves that affect the flow to the eye of the impeller. Also, there are variations in barometric pressure. Because of these variations between ideal conditions versus installed, a factor of safety or margin ratio is recommended. This is discussed in detail in American National Standards Institute/Hydraulic Institute (ANSI/HI) 9.6.1, "American National Standard for Centrifugal and Vertical Pumps for NPSH Margin." In general, it is recommended that at least 0.6 m (2 ft) be added to the $NPSH_R$ listed for the pump at a given flow, but this should be verified based on the application, type of pump, and the pump manufacturer's recommendations. Being more conservative may increase the cost of the structure for additional capacity, but also may benefit the long-term performance of the pumps.

TABLE 4.1 Atmospheric Absolute Pressure at Various Altitudes

Altitude Above Sea Level, m	Atmospheric Pressure, m of water at 20 °C	Altitude Above Sea Level, ft	Atmospheric Pressure, ft of water at 70 °C
0	10.3	0	34.0
300	10.0	1000	32.8
600	9.6	2000	31.6
900	9.3	3000	30.5
1200	9.0	4000	29.4
1500	8.6	5000	28.3
1800	8.3	6000	27.3
2100	8.0	7000	26.2
2400	7.7	8000	25.2
2700	7.4	9000	24.3
3000	7.1	10 000	23.4

TABLE 4.2 Vapor Pressure at Various Temperatures

Fluid Temp., °C	Vapor Pressure, m of water	Fluid Temp., °F	Vapor Pressure, ft of water
1.7	0.06	35	0.2
4.5	0.09	40	0.3
10	0.12	50	0.4
15	0.18	60	0.6
21	0.27	70	0.9
27	0.37	80	1.2

4.5 Pump Operating Range

The operating range of a pump is limited by hydraulic and mechanical considerations. Like most mechanical equipment, a pump operates best at its BEP, where flow is streamlined, reducing turbulence, erosion of parts, and hydraulic loads on bearings. Operating at their BEP, pumps can function for many years with minimum maintenance. Unfortunately, most wastewater pumps must operate over a range of pressures and flows at points considerably removed from the BEP.

Cavitation is a potential problem when a pump operates too far right on its head-capacity curve well beyond its BEP. At this point, the pump $NPSH_R$ increases rapidly as pump discharge increases.

If the pump operates to the far left on its head-capacity curve near shutoff head (zero flow), recirculation within the pump becomes a problem. At low flow, flow through the impeller is not smooth so that as the flow passes through the impeller, increasing the head, a portion of the flow reverses to return to the lower pressure at the eye of the impeller. This recirculation causes vibration in the pump and results in cavitation-like damage to the impeller. The flow at which this recirculation starts is closer to the BEP for high, specific-speed pumps than for low, specific-speed pumps.

Mechanically, pump thrust and radial loads increase as the flow moves away from the BEP. These increased forces reduce bearing life and increase shaft deflection that can, in turn, result in wear ring contact and failure of mechanical seals or packing and even the shaft.

For these reasons, it is good practice to limit the range of operation on the pump curve. The preferred operating range (POR) is a region of the pump curve defined in relation to BEP and represents the portion of the curve in which continuous operating conditions should be located, if possible.

The POR is defined by centrifugal pump standards such as the Hydraulic Institute Standards or American Petroleum Institute Standard 610 (API 610). The POR, as defined by Hydraulic Institute Standards or API 610, varies depending on the type of impeller, but, in most cases, extends from 70% to 120% of flow at the BEP. In some high-energy pumps, the POR range is tighter and may only extend from 80% to 115% of flow at BEP.

In contrast to the allowable operating range (AOR) is a manufacturer-defined region of the curve where the pump can be operated continuously without detrimental effects to the pump performance or equipment.

For systems that require pump operation over a wide range of flows, the pump manufacturer should be consulted for the recommended range of operation. Figure 4.9 shows an example of the POR and AOR ranges on a pump curve.

4.6 Affinity Laws

Affinity laws are used to predict centrifugal pump performance at different speeds and with different diameter impellers. These laws predict the slope of the head-capacity curve, the power required by a pump with great accuracy for changes in speed, and changes in impeller diameter (with less degree of accuracy).

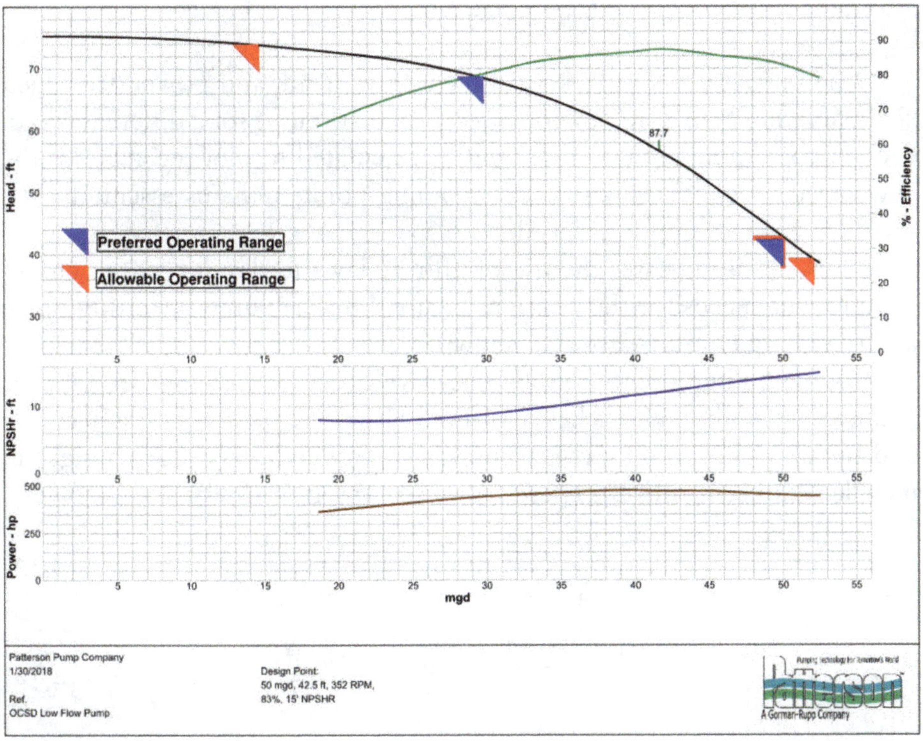

FIGURE 4.9 Typical Representation of Allowable Operating Range/Preferred Operating Range

For changes in speed, but with the impeller diameter remaining constant, the affinity law becomes

$$\frac{Q_2}{Q_1} = \frac{N_2}{N_1} \text{ or } Q_2 = Q_1 \frac{Q_2}{Q_1} \quad (4.10)$$

$$\frac{H_2}{H_1} = \frac{(N_2)^2}{(N_1)^2} \text{ or } H_2 = H_1 \frac{(N_2)^2}{(N_1)^2} \quad (4.11)$$

$$\frac{P_2}{P_1} = \frac{(N_2)^3}{(N_1)^3} \text{ or } P_2 = P_1 \frac{(N_2)^3}{(N_1)^3} \quad (4.12)$$

Where

Q_1 = pump flowrate, L/s (gpm) at speed 1;
Q_2 = pump flowrate, L/s (gpm) at speed 2;
N_1 = pump rotative speed, r/min (rpm);
N_2 = pump rotative speed, r/min (rpm);
H_1 = pump head, m (ft) at speed 1;
H_2 = pump head, m (ft) at speed 2;
P_1 = pump power required, kW (hp) at speed 1; and
P_2 = pump power required, kW (hp) at speed 2.

When using affinity laws, computed flow, head, and power are called *corresponding points*. To develop a head-capacity curve for a different speed, enough corresponding points should be calculated to plot the curve at the new speed. The efficiency of corresponding points remains essentially the same for changes in pump speed.

Figure 4.10 indicates the use of affinity laws to develop pump curves for different speeds for the pump in Figure 4.6. The pump operating range and operating envelope are also shown.

Corresponding points can also be developed for a change in impeller diameter when the speed is held constant. However, results are less accurate and the efficiency among corresponding points will vary. For changes in impeller diameter, with the speed held constant, the affinity law becomes

$$\frac{Q_2}{Q_1} = \frac{D_2}{D_1} \text{ or } Q_2 = Q_1 \frac{D_2}{D_1} \quad (4.13)$$

$$\frac{H_2}{H_1} = \frac{(D_2)^2}{(D_1)^2} \text{ or } H_2 = H_1 \frac{(D_2)^2}{(D_1)^2} \quad (4.14)$$

$$\frac{P_2}{P_1} = \frac{(D_2)^3}{(D_1)^3} \text{ or } P_2 = P_1 \frac{(D_2)^3}{(D_1)^3} \quad (4.15)$$

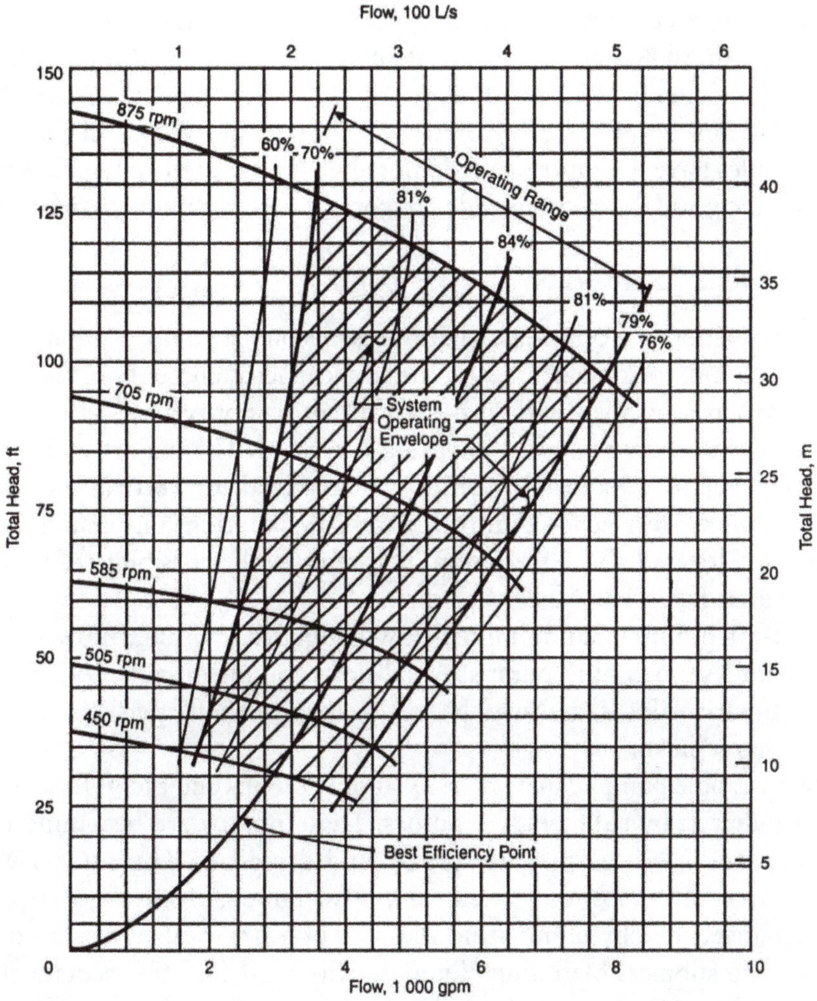

FIGURE 4.10 Pump Operating Range and Operating Envelope

Where

Q_1 = pump flowrate, L/s (gpm) at D_1;
Q_2 = pump flowrate, L/s (gpm) at D_2;
D_1 = impeller diameter, m (in.);
D_2 = impeller diameter, m (in.);
H_1 = pump head, m (ft) at D_1;
H_2 = pump head, m (ft) at D_2;
P_1 = pump power required, kW (hp) at D_1; and
P_2 = pump power required, kW (hp) at D_2.

The pump curve shown in Figure 4.6 illustrates the accuracy of using affinity laws on changes in diameter. For instance, the BEP for the 600-mm

(24-in.) diameter impeller in Figure 4.8 is 390 L/s (6200 gpm) at 35 m (110 ft) with 84% efficiency. The calculated BEP for the 480-mm (19-in.) diameter impeller is 315 L/s (5000 gpm) at 22 m (72 ft), but the curve shows the BEP at 300 L/s (4600 gpm) at 20 m (65 ft) with only 78% efficiency. The affinity rules have limitations and do not always accurately reflect changes in performance, and it is recommended to consult with the pump manufacturer.

4.7 Pump Construction

The construction of a centrifugal pump depends on the pump location (wet pit or dry pit) and, to some degree, on the size, capacity, and duty of the pump.

Pumps are installed with either a vertical or horizontal shaft configuration. Wet-pit pumps almost always are vertical. The majority of dry-pit stormwater and wastewater pumps are arranged for vertical mounting, which requires less space than horizontal installation and results in the drive being located above the pump. This reduces the possibility of flooding the drive in the event of a catastrophe.

Wet-pit pumps include mixed-flow propeller and axial-flow column pumps and submersible pumps. Axial-flow column pumps typically are driven by drives located above the wet well, with shafts that pass through the discharge head.

Submersible pumps can be both axial-flow pumps and radial-flow pumps that are driven by submersible motors. These pumps are becoming more common and are used in both wet-pit and dry-pit configurations. When used as wet-pit pumps, they are arranged for removal from the wet-pit for maintenance, thereby eliminating the cost of a dry well. When used in a dry pit, the submersible pump eliminates the need for intermediate shafts to connect a motor located above flood level with the pump and allows the station to continue in operation in the event of catastrophic flooding.

4.7.1 Radial-Flow Volute Pumps

The main parts of a radial-flow volute pump are the rotating assembly, the casing, and the bearing frame, or bearing housing. Pumps are classified as "close-coupled," "direct-connected," or "shaft-driven."

Close-coupled pumps have a rotating assembly and casing, but do not have a bearing frame. The impeller is mounted directly on the shaft of the motor and the casing of the pump is connected to the motor frame. Motor bearings serve as pump bearings. Close-coupled pumps such as solids handling submersible pumps are commonly used in municipal service.

Direct-connected pumps have a rotating assembly, a casing, and a bearing frame. The pump shaft is connected directly to the output shaft of the drive through a flexible coupling. The bearing frame of the pump contains the shaft bearings. Figure 4.11 shows a typical dry-pit, direct-connected wastewater pump.

The external shaft-driven, vertical, dry-pit pump is similar to the direct-connected pump except that the pump drive motor is located above and is connected to the pump by a drive shaft. Figure 4.12 shows a shaft-driven pump.

The rotating assembly includes the pump shaft, impeller, shaft sleeves, impeller wear rings, mechanical seals, bearings, and all pump parts, whether rotating or stationary, that must be assembled over the shaft.

When servicing a pump, the rotating assembly and the bearing frame typically are removed from the pump casing and taken to a shop where

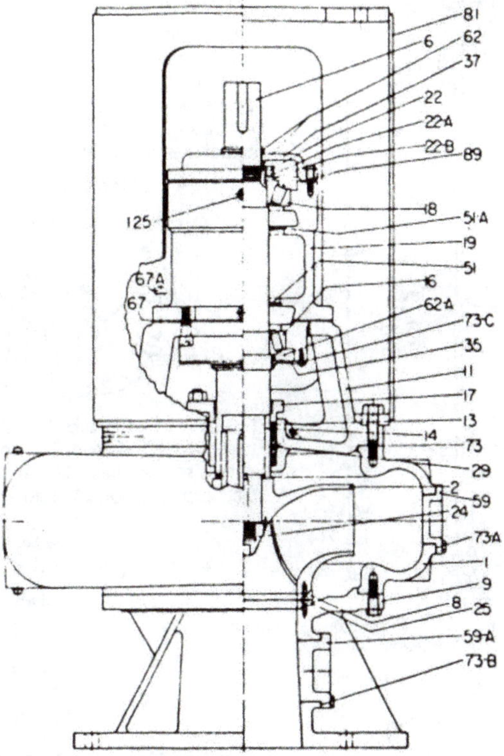

STATIONARY PARTS						ROTATING PARTS		
Ref. No.	No. of Pieces	Name of Part	Ref. No.	No. of Pieces	Name of Part	Ref. No.	No. of Pieces	Name of Part
1	1	Casing	51A	1	Grease Retainer-Outboard	2	1	Impeller
9	1	Suction Head	59	1	Handhole Cover-Casing	6	1	Shaft with Keys
11	1	Stuffing Box Head	59A	1	Handhole Cover-Suct. Hd.	8	1	Impeller Wrg. Ring
13	5	Packing Rings	67	1	Adjusting Shim Set	14	1	Shaft Sleeve
17	1	Gland	67A	2	Adjusting Bolt	16	1	Line Bearing
19	1	Bearing Frame	73	2	Ankorite Gasket	18	1	Thrust Bearing
25	1	Suction Head Wrg. Ring	73A	1	Rubber Gasket	22	1	Bearing Locknut
29	2	Seal Cage Halves	73B	1	Rubber Gasket	22A	1	Lockwasher
35	1	Line Brg. Cover	73C	1	Paper Gasket	22B	1	Bearing Washer
37	1	Thrust Brg. Cover	81	1	Motor Pedestal	24	1	Impeller Nut
51	1	Grease Retainer-Inboard	89	1	"O" Ring Seal	62	2	Seal Ring
			125	2	Alemite Fittings	62A	2	Seal Ring

FIGURE 4.11 Typical Vertical, Dry-Pit Volute Pump

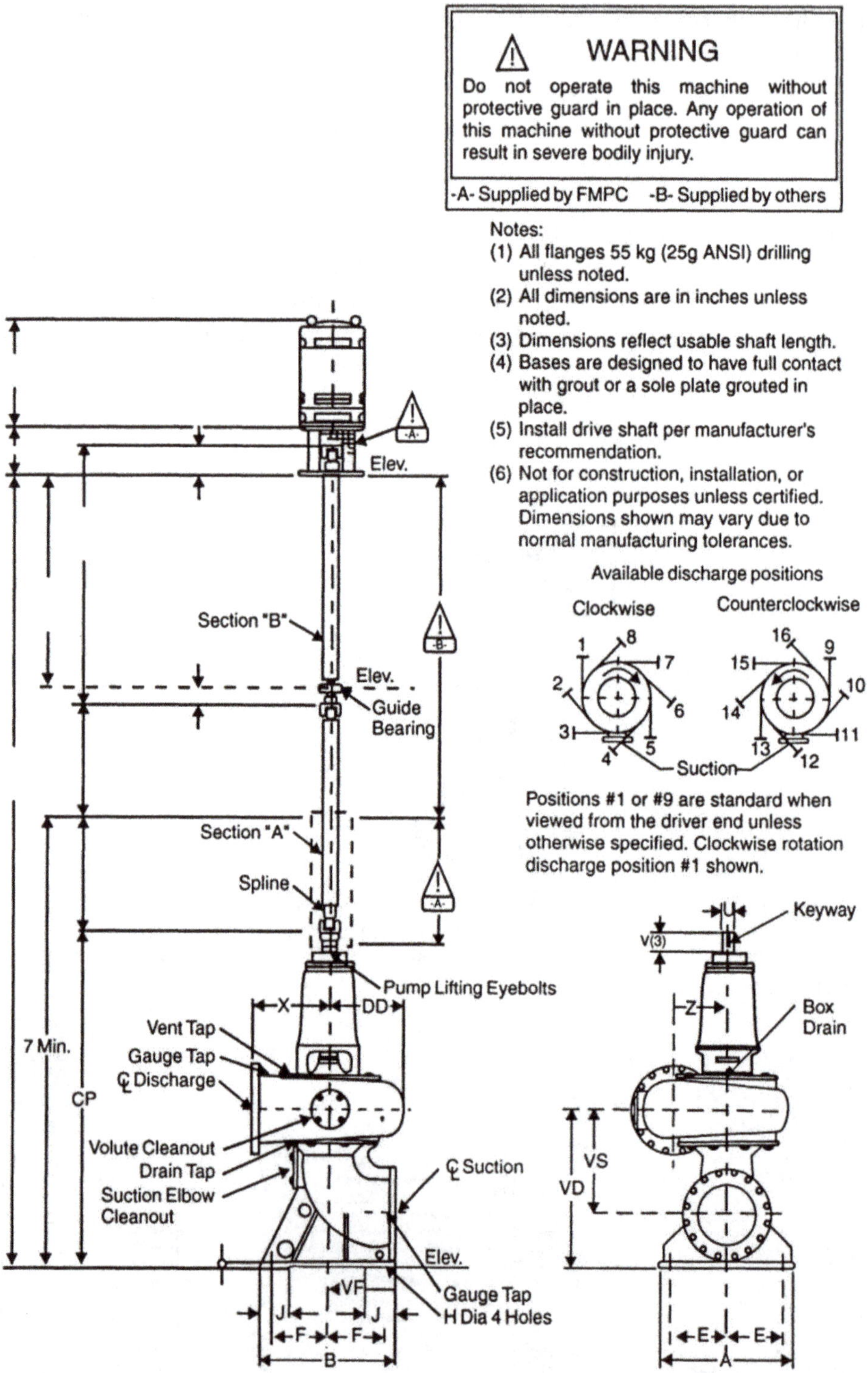

FIGURE 4.12 Typical Vertical, Shaft-Driven Volute Pump (in. × 25.4 = mm)

servicing is more convenient. For this reason, lifting devices to remove the rotating assembly and maneuvering space around the pump should be included in station design.

Replaceable wear rings typically are installed on the impeller and the casing at the inlet to the impeller to protect them from wear and reduce recirculation from the higher discharge pressure to the lower suction pressure. Because one rotates and the other is stationary, there must be clearance between the impeller and the casing, so there will be some recirculation. Abrasives in wastewater would erode the impeller and the casing without the wear rings, which should be replaced when clearances become excessive.

Wear rings can be designed for either close axial fit or close radial fit. Smaller pumps typically are furnished with close axial-running clearance, requiring adjustment of the clearance by axial adjustment of the impeller. Larger, more expensive pumps are furnished with radial or "L"-shaped rings having close radial clearance. The pump manufacturer should be consulted for recommended wear ring clearances.

The pump shaft transmits torque from the drive to the impeller and is supported by bearings mounted in the bearing frame. The shaft is subject to wear at the stuffing box, so replaceable shaft sleeves are installed on the shaft where it passes through the stuffing box to prevent damage to the shaft.

Where the drive is mounted above flood level, intermediate shafting is required to connect the drive to the pump. Depending on the distance, more than one section of shafting might be needed, and intermediate shafting must be selected carefully to eliminate destructive vibration. Vibration can be lateral (translational, movement from side to side, or up and down), or torsional (twisting).

All shafts have a natural frequency, or critical speed, that is dependent on length, tube dimensions, and tube material. The pump system produces lateral vibrations at several operating frequencies, known as *exciting frequencies*. Vibrations caused by imbalance occur at the pump operating speed; those caused by angular displacement in the shaft U-joint occur at 2 times the rotating speed; and vane-passing vibration occurs at the multiple of the rotative speed times the number of vanes in the impeller.

It is critical that exciting frequencies do not coincide with the critical speed of the shaft. The selection of shafting for a variable-speed pump is complicated because of the range of exciting frequencies produced over the speed range of the pump.

Until recently, most shafts were made of steel tubing. Shafts now are available in composite material of graphite and/or glass fibers in an epoxy thermoset plastic. These shafts are more expensive and difficult to handle than steel but have a higher critical speed and can be furnished in longer lengths.

The ANSI/HI 9.6.4 publication, "American National Standard for Rotodynamic Pumps for Vibration Measurements and Allowable Values," elaborates on this topic, and the pump manufacturer should be consulted during the design and testing of the pumps.

The pump casing is the stationary housing around the impeller and includes the volute, suction cover, and stuffing box cover. The volute casing is spiral shaped, collects the discharge from the impeller, and directs the flow to the discharge pipe. The suction cover and stuffing box cover are attached to the volute to enclose the impeller. One point in the volute, known as the *cut water*, is close to the outside diameter of the impeller. The flow area of the volute increases from the cut water to the discharge connection.

Hydraulic forces produce a radial thrust on the impeller proportional to the pump head and the projected area of the impeller. This thrust is a minimum at BEP but increases rapidly as the flow increases beyond or drops below the BEP and is maximum at shutoff head. Hydraulic Institute Standards provide data for approximating the radial loads on pumps.

Because of this thrust, a few large, high-head pumps are fitted with double volutes. In a double-volute pump, the volute is furnished with two cut waters 180 deg apart, with a vane from the second cut water to the discharge. This arrangement divides the flow between the two sides and the radial thrusts tend to cancel each other.

The suction cover connects the suction pipe to the pump casing. The suction cover can include a suction elbow and serve as a support base on some pumps.

Handholes are required on the suction pipe and volute to allow inspection of the pump and removal of solids lodged between the impeller vanes, clogging the pump. Handholes should be as large as possible, and the interior of the cover must match the contour of the pump.

The stuffing box cover on volute casings forms the back of the casing and contains a stuffing box to prevent leakage where the pump shaft passes through the cover.

The stuffing box can be sealed with packing or a mechanical seal. One or two rows of packing are in the stuffing box close to the impeller, with an adjacent lantern ring. Three or more rows of packing are located between the lantern ring and the outside of the stuffing box. The packing is held in the stuffing box by an adjustable gland that maintains pressure on the packing.

Because wastewater contains abrasives and packing requires lubrication, the packing must have an outside source of lubricant at a higher pressure (69 kPa [10 psi] or greater) than that in the pump. This ensures that leakage will be from the lubricant to the pump to keep abrasives out of the packing. This lubricant enters the stuffing box at the lantern ring, with a small amount of lubricant flowing into the pump casing and a small amount passing along the shaft to the atmosphere.

The most common lubricants are water or grease. If water is used, it must be from a clean source (if potable water, the source must be protected by an air break or be supplied through a backflow preventer, though some local codes do not allow backflow preventers). Figure 4.13 shows a typical packed stuffing box with water lubrication.

Spring-loaded or weight-loaded lubricators are available if grease lubrication is used. These lubricators should be controlled to provide lubricant only when the pump is operating.

If a packed stuffing box is used, the stuffing box will have a continuous, slight leakage, which should be piped to a drain.

Mechanical seals are also used to seal the stuffing box. A mechanical seal has a lapped rotating surface and a lapped stationary surface, and the smooth surfaces practically eliminate leakage. The faces are lubricated by a small amount of fluid that passes between the surfaces.

Mechanical seals are available in internal and external configurations and single and double seal arrangements. Because of abrasives in the

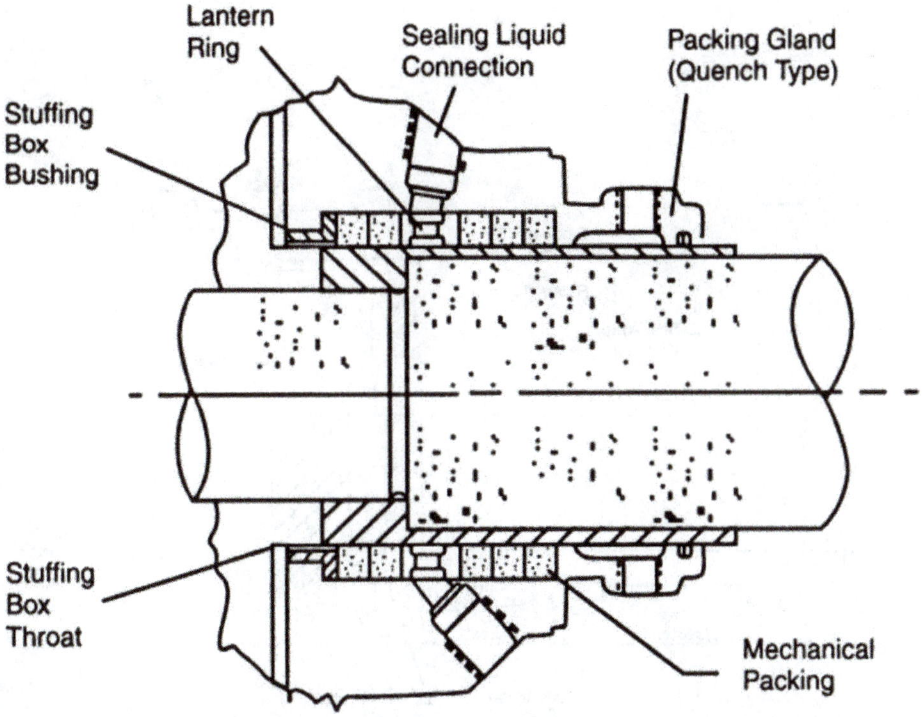

FIGURE 4.13 Stuffing Box With Packing

wastewater, double mechanical seals are recommended. Figure 4.14 shows a typical double mechanical seal.

Double mechanical seals consist of two seals arranged back-to-back, with one seal facing the impeller and the second facing the exterior of the pump. The space between the two seals must be provided with seal water to lubricate and cool the surface between sealing surfaces. Again, this lubricant must be at a higher pressure than the pumped wastewater. A small amount of this lubricant must flow through the seal to provide cooling, and this must be piped to a drain.

There are advantages and disadvantages to mechanical seals. Mechanical seal failures are complete and result in considerable leakage until the pump is shut down. The packed stuffing box leakage gradually increases as the packing wears.

The replacement of a conventional mechanical seal requires the disassembly of the pump to place the mechanical seal over the shaft of the pump. Packing can be replaced while the pump is assembled. Split mechanical seals that allow seal replacement without pump disassembly are a recent development but are expensive and require more space. The pump manufacturer should be consulted to ensure that the pump can be fitted with split mechanical seals.

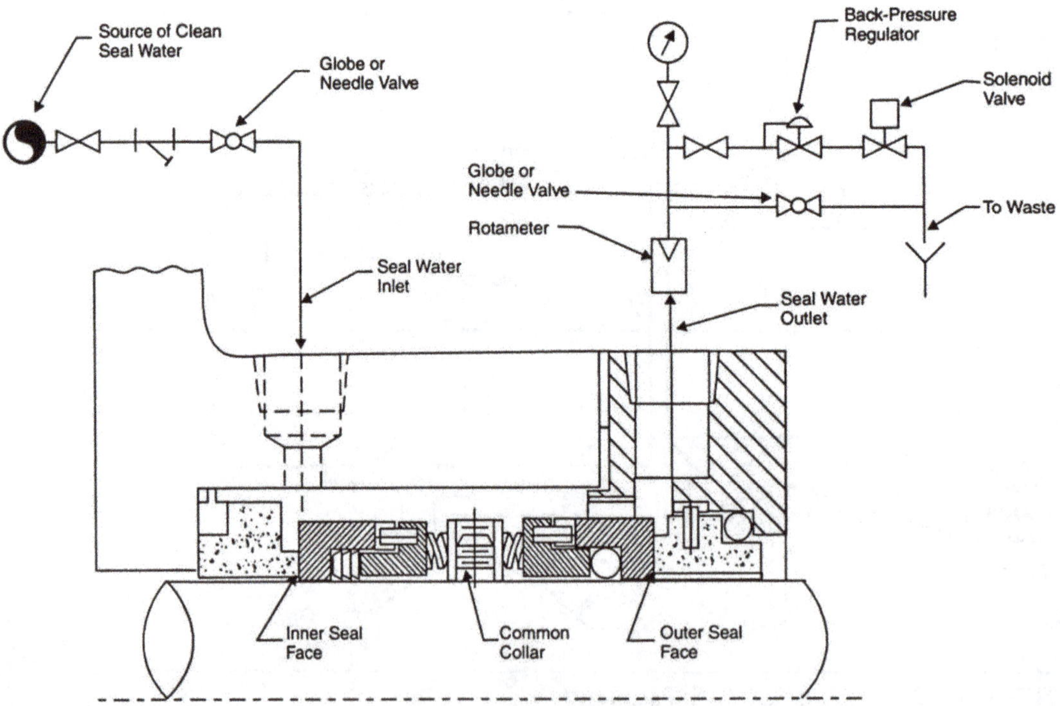

FIGURE 4.14 Typical Double Mechanical Seal

The bearing frame on direct-connected and shaft-driven pumps provides support for the pump shaft and contains the pump bearings.

Hydraulic forces impose both radial- and axial-thrust loads on the impeller. The discharge pressure acts on both sides of an enclosed impeller. These loads cancel each other, except at the eye of the impeller, where one side of the impeller is acted on by the suction pressure while the other side sees the discharge pressure. This imbalance produces thrust.

In large, high-head pumps, the axial thrust can be reduced by installing pump-out vanes on the back of the impeller or installing a second set of wear rings on the back of the impeller, opposite the edge of its eye, and drilling a balancing port from the inlet to the chamber inside the back wear rings. This allows the suction pressure to act on both sides of the impeller, reducing axial thrust.

The outboard bearing, the one farthest from the impeller, is designed for both axial- and radial-thrust loads. The inboard bearing is designed for radial loads.

Most new volute pumps have anti-friction-type ball or roller bearings, with the roller bearings only used on larger pumps with heavier loads. These antifriction bearings can be lubricated with oil or, more commonly, grease.

The thrust load of some large (1-m [35-in.]) and larger high-head pumps can exceed the load capacity of antifriction bearings. For these applications, a Kingsbury-type thrust bearing can be mounted in the pump or motor.

4.7.2 Axial-Flow Column Pumps

Axial-flow, wet-pit column pumps differ from radial-flow pumps in many ways. The main parts of an axial-flow, wet-pit pump include the rotating assembly, suction bell, discharge bowl or bowls, column, and discharge head and drive.

The impellers are high, specific-speed, mixed-flow, or propeller-type impellers. Mixed-flow impellers can be semi-open or enclosed and discharge flows at an angle from the shaft. The propeller impeller is an open-type impeller and discharges flow parallel to the shaft.

These pumps can be single-stage or multistage as necessitated by the system discharge head.

The suction bell directs the flow to the inlet of the impeller and houses the shaft bottom-bearing of the pump.

The discharge bowl encloses the impeller and contains the pump bearings. Discharge bowls for mixed-flow impellers are provided with diffuser vanes that collect the flow from the impeller and guide it axially up the column or to the inlet of the next-stage impeller.

The column connects the discharge bowls to the pump discharge head, which can be above or below the floor that supports the pump in the wet well. The discharge head diverts the flow from parallel to the pump shaft to perpendicular to the shaft for connection to external piping.

The pump shaft passes through a stuffing box in the discharge head and passes through the column pipe.

Axial-flow, wet-pit pumps typically are furnished with sleeve bearings designed for radial loads only. Axial thrust is absorbed by a thrust bearing in the drive and, because the pumped fluid contains grit, the shaft and bearings are mounted in a shaft-enclosing tube. To lubricate the sleeve bearings, clean water or oil is admitted through the top of this tube or grease is supplied to the bearings through grease tubes. The enclosing tube is structurally stabilized in the column pipe to maintain alignment.

4.7.3 Construction Materials

Considerations in selecting construction materials include the fluid being pumped, abrasive material in the fluid, and cavitation potential. Raw wastewater and stormwater are neutral fluids (at a pH of 7) at temperatures between 5 and 30 °C (40 and 85 °F). However, wastewater and stormwater do contain abrasive materials (grit) that cause wear in pumps.

The casing and bearing frame must be of heavy construction to absorb the force of the fluid being pumped and the loads imposed on the bearings.

Almost without exception, the casing and bearing frame are made of cast iron. Higher-strength ductile iron sometimes is used for larger, higher-head pumps. Cast steel occasionally is used for the casing in high-head pumps on flows containing large quantities of grit to reduce erosion in the casing.

The shaft furnished in most pumps is of carbon steel, which has the strength needed to deliver the input power to the impeller.

Sleeves are installed on the shaft where it passes through the stuffing box, protecting the shaft from abrasive wear. Shaft sleeves should be hardened to 350 to 450 Brinell Hardness (BNH), and those on packed pumps are typically of 400 series stainless steel or its equivalent casting alloy. This material is resistant to wear and is machinable. Shaft sleeves of pumps with mechanical seals are not exposed to the abrasive wear of a packed pump and can be of 300 series stainless steel.

Impeller material considerations include the quantity of grit in the wastewater being pumped, the flowrate in relation to best efficiency, and the cost of the material.

Most wastewater service impellers are made of cast iron, which is the least expensive material and provides adequate service where the grit load is not excessive.

For large, higher-head pumps, the extra cost of materials more resistant to erosion and cavitation damage is justifiable. Table 4.3 contains information

TABLE 4.3 Material Resistance to Abrasion and Cavitation Damage

Cavitation Resistance		Abrasion Resistance	
1	Cast iron	1	Cast iron
2	Bronze	2	Bronze
3	Cast steel	3	Manganese bronze
4	Manganese bronze	4	Nickel/aluminum/bronze
5	Monel	5	Cast steel
6	400 series stainless steel	6	300 series stainless steel
7	300 series stainless steel	7	400 series stainless steel
8	Nickel/aluminum/bronze		

Note. Listed in order of increasing resistance (cast iron least resistant)

from *Pump Handbook* (Karassik et al., 2007), which lists materials in order of increasing resistance to cavitation and abrasive wear damage. In each case, cast iron displays the least resistance. The extra cost of higher-resistant materials must be weighed against the benefits of longer impeller life.

Wear rings are installed to reduce recirculation in a pump and are subject to erosion and galling.

Stainless steel of the 400 series or its equivalent casting alloy typically is used for the aforementioned parts. The impeller wear ring should be hardened to 250 to 350 BNH and the casing wear ring should be hardened to approximately 100 BNH harder than the impeller ring to reduce the potential for galling.

5.0 POSITIVE DISPLACEMENT PUMPS

Most positive displacement pumps have limited application in stormwater and wastewater pumping because of close clearances and wear on the pump parts. The two types of positive displacement pumps commonly used are the Archimedes screw pump and the pneumatic ejector, which are at the opposite ends of flowrates. The screw pump is typically used for large flows and the pneumatic ejector is typically used for small flows.

Two other significant types of pumps are the airlift pump and the progressive cavity pump.

5.1 Archimedes Screw Pumps

One of the oldest types of pumps is based on the Archimedes screw principle, in which a revolving spiral, rotating in an inclined trough, pushes the flow up the trough.

Screw pumps are lift pumps that draw water from an open channel, lift it, and discharge it to another open channel at a higher elevation. The screw pump has two distinct advantages when pumping stormwater or wastewater: (1) it can pump large solids and stringy material without clogging and (2) it pumps variable flowrates over a wide range of flows without complicated control equipment, thereby matching incoming flow.

Pumps are driven by constant-speed motors through a gear box at speeds typically between 25 and 75 r/min (rpm), and screw pump drive motors by design do not overload. The capacity of the screw pump increases as the flow level entering the screw rises until it reaches the fill point, which is the approximate center of the screw. Though pump capacity cannot increase as flow level rises above the fill point, the pump head decreases, reducing pump horsepower. If the drive motor is sized for the pump capacity at the fill point, the motor cannot be overloaded.

Screw pumps are either open or enclosed. Open screw pumps have screw flights welded to the outside of a central torque tube. Figure 4.15 shows an open screw pump installation.

Open screw pumps are available with diameters from 0.6 to 3.6 m (24 to 144 in.) and capacity ranges from 60 to 4700 L/s (1000 to 75 000 gpm). Because of structural limitations of the torque tube, the lift of the open screw pump typically is limited to 7.5 to 9 m (25 to 30 ft), but with special construction can be extended to 12 m (40 ft).

Standard angles of incline of an open screw pump are 30 and 38 deg. A pump installed at 30 deg will pump more than one installed at 38 deg but will require more space and have less available lift.

The torque tube is supported by bearings at the bottom and the top of the pump, and the drive motor and gear box are located at the top of

FIGURE 4.15 Open Screw Pump Installation

the pump above the discharge point. The bottom bearing typically is of the grease-lubricated sleeve type and, because it is submerged, should be lubricated automatically. The top bearing is designed for both thrust and radial loads.

Open screw pumps require close clearance between the perimeter of the screw and the trough to maintain pump efficiency. When fill concrete is poured in the trough, screed boards are attached to the flight to provide the correct clearance.

Because there is clearance between the flights and the trough, the pump is self-draining. This feature reduces starting loads and eliminates problems with freezing but reduces pump efficiency. Typical shaft efficiencies of open screw pumps are 75% at maximum capacity and 65% at 25% of capacity.

The open, rotating flights of the open screw pump present a safety hazard, and it is strongly recommended that a grating or cover be installed over the pump trough. In warm climates, sunscreens may be required to prevent idle pumps from bowing from extreme temperatures.

Enclosed screw pumps have conically pitched helical flights welded inside an exterior steel cylinder. They are available with diameters from 0.6 to 2.75 m (24 to 110 in.) and capacities from 60 to 2200 L/s (1000 to 35 000 gpm). Standard angles of incline are 38 and 45 deg. A pump installed at 38 deg will pump more than a pump installed at 45 deg.

The upper bearing of the pump is located above the discharge point and is designed for thrust and radial loads. The lower pump cylinder is supported by steel-reinforcing rings that ride on roller bearings. The lowest bearing is located above the pumped fluid, with the inlet to the pump cantilevered so that the bottom bearing is not submerged.

Unlike the open screw pump, the lift of an enclosed screw pump is not restricted by the structural limitations of the torque tube. For lifts of more than 7.5 m (25 ft), additional roller bearings can be mounted on the cylinder to extend the lift.

The shaft efficiency of the enclosed screw pump is approximately 90% for flows between 50% and 100% of maximum flow and approximately 80% at 25% of maximum flow. The reason for the high efficiency is the elimination of the recycle flow of the open screw pump. The only losses in an enclosed shaft pump are bearing losses and gear box losses.

One disadvantage is that the enclosed screw impeller is not self-draining, increasing the starting torque of the motor, and, in cold climates, causing problems with freezing in the cylinder. To eliminate this problem, a clutch can be installed in the gear box to allow the pump, after being stopped, to go into reverse rotation to drain. The clutch limits the speed of the reverse rotation. Motion detectors are needed to prevent restarting while the pump is in reverse rotation.

Another potential problem with an enclosed screw pump is the danger of air binding. If the water level at the inlet to the pump rises above the top of the exterior steel cylinder, air is trapped in the pump, severely reducing pump capacity. To prevent air binding, inlet level controls should be installed to sound an alarm and start standby pumps before the wet well level reaches the crown of the pump cylinder.

Enclosed screw construction is simplified because the trough required by the open screw pump is eliminated.

5.2 Pneumatic Ejectors

Pneumatic ejectors are used where the flow of wastewater to be pumped is less than the capacity of a small centrifugal pump. The typical capacity of a pneumatic ejector is 20 L/s (320 gpm) or less.

A typical pneumatic ejector, as shown in Figure 4.16, consists of a receiver, called a *pot*; an air compressor; controls; and associated piping. Check valves are located on the inlet and outlet of the pot and a vent line is connected to the top of the pot. A three-way solenoid valve is installed on the vent line. In one position, the pot is vented to the atmosphere and wastewater can flow in through the inlet valve. When the pot is full, the controls reverse the solenoid valve, admitting compressed air into the pot and forcing the wastewater out through the discharge check valve. When the pot empties, the controls reverse again, venting the pot.

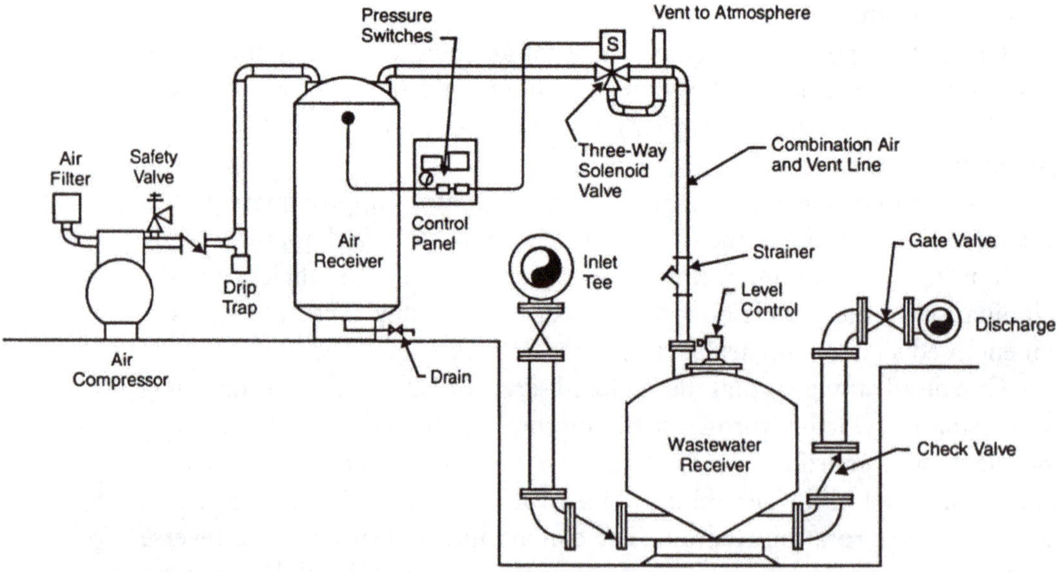

FIGURE 4.16 Pneumatic Ejector and Associated Piping

Because the pot is filled by gravity flow, the pot must be several feet below the inlet sewer to provide sufficient head to overcome the friction loss in the inlet pipe and check valve.

The pot typically is sized for a 1-minute cycle and the holding capacity of the pot is numerically equal to the ejector flowrate. For instance, a 6-L/s (100-gpm) ejector would have a 360-L (100-gal) capacity pot. For a 1-minute cycle, the pot fills for 30 seconds and then ejects in 30 seconds. Therefore, the inlet piping and the discharge piping and force main must be sized for 200% of the ejector rating. The instantaneous discharge rate of the 6-L/s (100-gpm) ejector would be 13 L/s (200 gpm).

Pneumatic ejectors typically are furnished in pairs so that one pot is filling while the other is ejecting. Ejectors are available with steel or cast-iron pots. Cast iron reduces corrosion and is recommended for municipal service.

5.3 Airlift Pumps

Airlift pumps are seldom used in wastewater pumping stations because their total lift is 1.75 m (6 ft) or less and their efficiency is merely 20% to 30%. However, airlift pumps are virtually uncloggable.

The airlift consists of a vertical riser pipe submerged in a wet well with an air diffuser at the bottom. Compressed air is discharged through the diffuser to mix with the wastewater. Because the air and water mixture density are less than the density of the surrounding wastewater, atmospheric pressure forces the mixture to a higher head.

5.4 Progressive Cavity Pumps

Progressive cavity pumps, by themselves, typically are not used for wastewater or stormwater pumping, but are common at WRRFs for pumping heavy sludges.

Progressive cavity pumps can be used in grinder pump systems. These are small-capacity pumps typically used in remote, residential collection systems. Before entering the pump, wastewater passes through a grinder to reduce solids to a size that will pass through the pump and the small-diameter force main.

6.0 FORCE MAINS

Two conflicting requirements are factors in selecting force main size. Velocity in the force main should be kept high to ensure the transport of grit and other solids, yet low velocity best reduces friction loss, energy costs, and water hammer.

Chapter 5 presents a detailed discussion of the requirements for sizing force mains, calculating force main friction losses, and water-hammer considerations.

7.0 WET WELLS

Wet wells store wastewater before pumping, provide submergence over the pump suction, and, most importantly, provide a smooth transition of flow to the pump suction. Chapter 3 discusses design considerations for the wet well.

8.0 PUMP DRIVES

Almost all pumps are driven by either electric motors or internal combustion engines. The most important criterion for selecting the type of drive for stormwater and wastewater pumping stations is reliability. The electric motor is more reliable and requires less maintenance than an engine drive.

Where reliable sources of power exist, electric motors are preferred. Ensuring reliable power is good practice and often requires that power be obtained from two independent power sources.

Where dual power sources are not available, the best choice is direct drive of the pumps by engines or the installation of a standby engine generator. The engine generator is advantageous in that power can be furnished to any of the installed pumps and the control system and station auxiliary systems, such as lights, heating, and ventilation.

Other considerations in pump drive selection include drive costs, operating costs, continuous versus intermittent operation, the need for variable-capacity operation, and the need for remote or automatic operation.

When evaluating the cost of the drive, expenses in addition to the first cost of the equipment must be included. These include the cost of controls, conduit, wiring, and starting equipment for motors and additional building space, extra floor strength, fuel system, jacket water cooling system, gear drives, and material-handling equipment for engines.

Operating costs include energy or fuel maintenance material and personnel expenses and the cost of operating auxiliary systems.

The duty cycle of the station can affect the energy costs of a pumping station. Electrical power charges typically have two components: the energy charge and the demand charge. The energy charge is based on total energy used, and the demand charge is based on highest use of power over a given period of time. The cost of fuel used in engines does not have a demand charge, favoring their use at stormwater pumping stations, which operate intermittently so use little energy. When operating, however, their energy demands are high.

An engine does not have restrictions on time between starts, but the frequency of starting a motor is restricted because of the heat generated by the inrush current when starting. Chapter 3 discusses the time required between starts.

Engines are designed for variable-speed operation. Electric motors are constant-speed drives and require additional controls or equipment to adapt them for variable-speed operation. Electric motors can easily be adapted to automatic control and remote operation. The automatic control of an engine is more complicated because it is monitoring from a remote location because of engine auxiliary systems.

8.1 Electric Motors

The following is an overview of motor drive requirements when applied to pumping systems. Electric motors are designed for either alternating current or direct current. Direct current motors seldom are used in municipal systems with the wide availability of alternating current power.

Alternating-current motors can be either single phase or three phase. The motor rating of single-phase motors is limited to approximately 7.5 kW (10 hp), so almost all motors used in pumping stations are three-phase, alternating-current motors.

The three types of alternating-current motors are the squirrel cage induction motor, the synchronous motor, and the wound-rotor induction motor. The speeds of the synchronous motor and the squirrel cage induction motor are constant. Wound-rotor induction motors also operate at a constant speed but can be made to operate at different speeds by varying the resistance in the motor secondary circuit.

The synchronous speed of a motor is calculated by the following equation:

$$\text{r/min (rpm)} = \frac{120 \times \text{frequency}}{\text{Number of motor poles}} \tag{4.16}$$

Synchronous motors operate at this speed, but squirrel cage and wound-rotor induction motors at full speed have a 2% to 3% slip loss and, therefore, operate at 97% to 98% of the synchronous speed.

The speed of stormwater and wastewater pumps should not exceed 1800 r/min (rpm), and 1800-r/min pumps should be used only when the system head requires a pump of this speed. Most stormwater and wastewater pumps operate in the range of 600 to 1200 r/min to reduce wear on the pump.

Higher speed pumps are less expensive, but lower speed pumps are larger, more durable, and less susceptible to wear and vibration. As pump size increases, pump speed decreases. Pump head is proportional to the square of the tip speed of the impeller. A larger diameter impeller will rotate at a lower speed to develop the same tip speed of a smaller impeller operating at a higher speed.

Squirrel cage induction motors are used most often because they are low in initial cost, require little maintenance, and have a proven record of reliability. However, at speeds below 600 r/min, squirrel cage induction motors have low power factors, which may increase power costs.

Synchronous motors are more expensive than squirrel cage induction motors and are typically used only in larger sizes (400 kW [530 hp]) and at lower speeds. However, because the synchronous motor can operate at unity or a leading power factor, it can be used in systems where power factor correction is required or it is beneficial to reduce power costs.

Wound-rotor motors are more expensive than squirrel cage induction motors and require more maintenance because of the brushes on the motor slip rings but can be made to operate at speeds below full speed.

Pump-system operation is often improved, and energy costs are reduced if two-speed motors or step-speed control of wound-rotor motors is used.

The wound-rotor motor operates at peak efficiency only when operating at its maximum speed. At reduced speeds, the motor is subject to slip losses generated by the resistance in the wound-rotor motor secondary circuit. Slip losses can be calculated by the following formula:

$$\text{Slip loss} = \text{Pump power} \times \frac{(\text{Full speed} - \text{Reduced speed})}{\text{Reduced speed}} \tag{4.17}$$

Where

Slip loss = power lost, kW (hp); and
Pump power = pump shaft power at the reduced speed, kW (hp).

The total power supplied by the motor is the pump shaft input power plus the slip loss. The energy in the slip loss is typically released in the form of heat in a resistance bank, which in cold climates can be used to heat the building. In warmer climates, or during warm periods, this heat must be wasted.

An additional advantage of the step-speed control of the wound-rotor motor is its ability to use the steps to bring a motor up to full speed slowly. This "soft start" eliminates the high inrush current of a squirrel cage induction motor and reduces the required time between starts.

Squirrel cage induction motors are available for operation at two different speeds. However, unlike the wound-rotor motor that can be adjusted to operate at any speed, the lower speed of the two-speed induction motor must be a synchronous speed less slip loss. A two-speed squirrel cage induction motor can be a two-speed, two-winding motor or two-speed, one-winding motor.

The ratio between speeds of a two-speed, two-winding motor can be the ratio between any two synchronous speeds. The reduced speed of a two-speed, one-winding motor must be 50% of the higher speed. Unlike the wound-rotor motor that loses efficiency at reduced speed, the two-speed squirrel cage induction motor maintains high efficiency at both speeds because the two-speed motor is, in fact, two separate motors.

8.2 Variable-Speed Motor Operation

Pump systems that must match inflow to the station are typically furnished with variable-speed pumps. Electric motors are constant-speed drives. There are, however, ways to provide stepless, variable-speed control with electric motor drives.

8.2.1 Variable-Frequency Drive

The variable-frequency drive is the most common method of variable-speed drive, replacing the older forms of variable-speed drives. This drive has been commonly used since the early 1980s and is an efficient, reliable method for variable-speed operation of squirrel cage induction motors and, in larger sizes, synchronous motors.

The variable-frequency drive takes alternating current, converts the power to direct current, and then reconverts the direct current to alternating current at a different frequency. Because the synchronous speed of a motor is proportional to frequency, the speed of the motor depends on the output frequency of the variable-frequency drive controller. Because the motor is operating at its full synchronous speed, there are no slip losses in the control.

There are, however, losses in the controller for converting to direct current and reconverting to alternating current. These losses are fairly constant over the full speed range of the controller. Therefore, the variable-frequency drive has poorer efficiency between 95% and 100% of full speed than other variable-speed drives that have slip losses, but the efficiency at less than 95% speed is much better than other drives.

There are several types of variable-frequency drives, with the pulse width modulation and current source converters being the most prevalent. The output power produced by the variable-frequency drive is not a smooth sinusoidal wave and induces harmonic disturbances and a low-power factor on the power supply source. Adding reactance to the circuit can help, but does not eliminate the problem. For this reason, motors often are sized to operate at 80% to 90% of their rated power and are furnished with a 1.15 service factor.

Care must be taken in the selection of engine-generator sets for use with variable-frequency drives. In addition to the 5% power loss in the converter, some oversizing is required to compensate for the harmonics in the power supply, especially if two or more variable-frequency drives are powered by the generator.

Variable-frequency drives can be furnished to operate several motors in parallel. This reduces the initial cost of the variable-speed drive, but is not a recommended practice because of the need for reliability. If the variable-frequency controller malfunctions, it would shut down too large a portion of the pumping station. A separate variable-speed controller is recommended for each variable-speed pump.

Other methods of producing stepless, variable-speed electric drives are available in two basic forms: (1) where the output speed of the motor is variable and (2) where the output speed of the motor is constant but operates through a clutch that produces a variable-output speed.

8.2.2 Wound-Rotor Motor Variable Speed

The output speed of wound-rotor motors can be made stepless if the resistance is infinitely variable.

A solid-state controller is available that converts the energy in the slip losses of the secondary circuit to direct current, reconverts the direct current to alternating current, and feeds this power back into the power supply to the drive. This control is known as *wound-rotor regenerative secondary control.*

The regenerative secondary controller efficiency approaches that of variable-frequency drives in that the energy in the slip losses is converted to useful work. However, the regenerative secondary controller has not been popular because of the additional cost and maintenance requirements of wound-rotor motors. Regenerative secondary controllers do, however, have energy-saving potential on existing systems with wound-rotor motors.

Liquid rheostats have been used for many years with wound-rotor motors to provide stepless-speed control. In these controls, the resistance in the secondary control is varied by changing the area of electrodes covered by an electrolyte.

It is important that electrolyte strengths be maintained equally in rheostats that control pumps that operate in parallel. If electrolyte strengths differ, pump speeds will be different because, even though the submerged area is equal, the secondary resistance will be different.

The energy of the slip losses heats the electrolyte, which then must be cooled. This heat can be used to heat the building by passing the electrolyte through a heating coil, or the heat can be transmitted to the fluid being pumped to liquid heat exchangers on the pump suction or discharge piping or can be exhausted to the outdoors.

8.2.3 *Variable-Voltage Motors*

Variable-voltage motors are specially designed high-slip squirrel cage induction motors. The primary voltage to these motors is varied by saturable reactors or silicon-controlled rectifiers to vary the output speed of the motor.

This type of control is not recommended for municipal service because it is inefficient and slip losses generate heat in the motor, causing it to run hot.

8.2.4 *Variable-Speed Clutches*

Variable-speed clutches have been available for years but are less common today because of the development of the variable-frequency drive. Clutches are available in magnetic drive and fluid drive. A squirrel cage induction motor most often is used to drive both clutches.

The magnetic drive, or eddy-current coupling, consists of a steel ring member, or drum, driven at constant speed by the output shaft of the motor and magnetic coils mounted on the output shaft of the drive. The strength of the magnetic field of the coils is varied to adjust the slip speed between the input and output shafts. These drives generate slip losses, with the lost energy calculated by Equation 4.17. In addition to the slip losses, bearing and windage loss must be included in calculating total energy loss.

Most magnetic drives are air cooled with heat generated by the losses exhausted to the space around the drive. This heat can be used for space heating, but during warm weather it must be exhausted. Large units can be water cooled.

Hydraulic drives are similar to magnetic drives, with the slip between input and output shafts controlled by the viscous shear of oil that separates the drive and the driven clutch plates.

Other types of variable-speed drives, such as adjustable belt pulleys and shift gears, are not used in wastewater and stormwater pumping stations. Figure 4.17 shows the relative efficiencies of the various types of variable-speed drives.

8.3 Internal Combustion Engines

Internal combustion engines can be fueled by gasoline, diesel oil, natural gas, liquid petroleum gas, or, if available, sludge gas. Engines can be spark

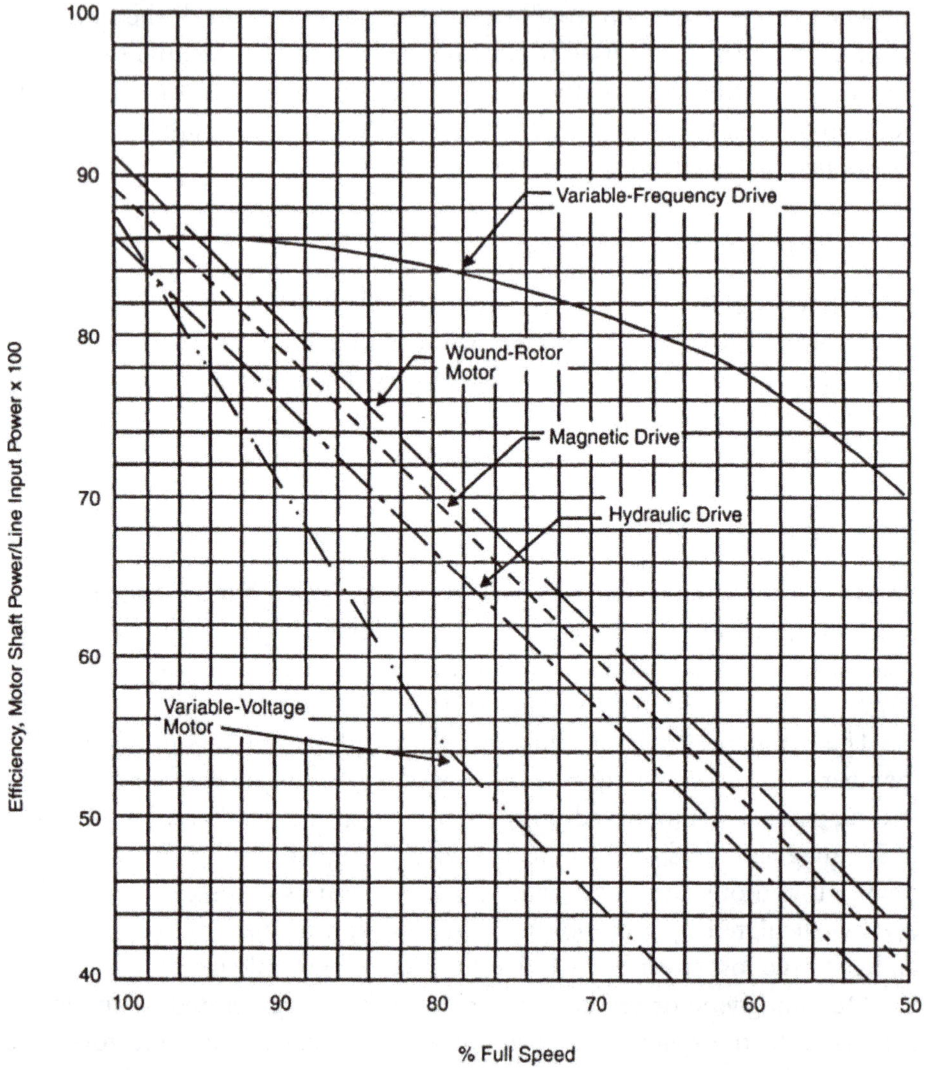

FIGURE 4.17 Variable-Speed Drive Efficiencies

Note. Approximate efficiencies; actual efficiency will vary with size and speed of motor and system curve characteristics

ignited, with the fuel ignited by a spark plug, or diesel engine, with the fuel ignited by the heat of compression inside the cylinder. Dual fuel engines are available with gas and diesel oil that are ignited by the heat of compression of the diesel fuel.

Most engine drives include a gear box between the engine and pump to allow both pieces of equipment to operate at their most efficient speed. Right-angle gears are used where the engine is horizontal, and the pump is vertical.

Low-speed, spark-ignited gas or diesel engines are most often used and, with proper maintenance, will operate for years. High-speed gasoline engines are the least expensive but have higher maintenance costs. Also, because gasoline is more volatile than other fuels, safety is a problem and difficulties can arise with the fuel if it is stored for an appreciable time.

Engines can be naturally aspirated or pressure charged. The volume of the air-fuel mixture that can be introduced to the cylinders in a naturally aspirated engine is limited, so the power generated also is limited. In a pressure-charged engine, the combustion air is pressurized by an exhaust-driven or power-train-driven blower to compress the combustion air. This forces a larger volume of air-fuel mixture into the cylinders and increases the power and efficiency of the engine. Also, the exhaust gas from a pressure-charged engine tends to be cleaner than the exhaust gas from a naturally aspirated engine.

One advantage of engine drive is that the engine does have the capability of a variable-speed drive. Occasionally, an engine can be used to provide standby drive for a pump by mounting the motor on the top of a dual right-angle gear drive with the motor shaft connected to the gear output shaft and the engine connected to the horizontal gear input shaft through a clutch that disconnects the engine when the motor is used. When the engine drives, the motor rotor windmills.

Auxiliary systems required for internal combustion engines include the fuel system, lubrication system, jacket water cooling system, and exhaust system.

Stations designed for continuous operation that use gasoline or diesel fuel must have sufficient storage space to receive the full capacity of delivery trucks plus several days' reserve in the event delivery is delayed.

Liquid petroleum gas (LPG) occasionally is used in small systems or where an engine is used for standby power. Though stored under pressure as a liquid, LPG is used in the engine as a gas. It can be withdrawn from storage as a liquid or a gas, but, if large quantities are withdrawn as a gas, the heat of vaporization cools the liquid to prevent further gas withdrawal. Therefore, LPG is typically withdrawn as a liquid and vaporized to a gas in a vaporizer, using heat available in the jacket water.

Natural gas typically is piped to the station from a gas distribution system, which should have a reliable gas supply and adequate pressure. Turbo-charged and intercooled engines require gas at high pressures (170 to 210 kPa [25 to 30 psi]) but are more efficient than naturally aspirated engines.

Only 25% to 35% of the energy supplied to an engine is converted to shaft output power. The balance of the heat is transmitted to the jacket water and the exhaust gas, with a small amount (6% to 10%) radiated to the surrounding space.

Twenty to thirty percent of the engine input energy is transmitted by the engine to the jacket water. The jacket water leaves the engine at 80 to 90 °C (176 to 194 °F) and must be returned to the engine 8 to 10 °C (46 to 50 °F) cooler to cool the engine. If there is a need for waste heat, such as building heating, this is an ideal source of heat. If the heat is not required, it must be dissipated. Air-cooled radiators are most often used in smaller systems. Larger systems can use remote, mounted fluid coolers or cooling towers with jacket water-heat exchangers. Water from a cooling tower should not be used directly in the cooling system or solids in the cooling tower water might plug small openings in the engine.

Additional waste heat can be recovered from the exhaust gas using heat recovery silencers. Exhaust silencers are required on all engines to prevent noise pollution.

9.0 PUMP SELECTION

Before selecting pumps, the design engineer must evaluate the system characteristics and determine the requirements for the pumping systems, which will vary with each system.

9.1 System Design Considerations

Pumps installed in a station must be capable of discharging the peak design flow at the system head required with aged pipe. It is normal practice, and typically required by code, to provide this pumping capacity with the largest pump out of service. In large stations, especially those with internal combustion engine drives, it may be desirable to discharge the design peak flow with one standby pump and one out of service for maintenance.

Pump selection evaluation should consider flowrates, system head-capacity curves, station location and area served, and force main velocities.

9.1.1 Flowrates

The system must be designed to discharge the design peak flow, which is the maximum flow the station will be required to discharge during the design life of the station. This flow may not develop until the end of the design life of the station, and the pumps will operate at this flow for only short periods of time.

Because the design flowrate is only occasionally achieved, other flowrates are more important to the design engineer. The station should be designed to operate efficiently during initial, interim, and design year average flows.

Interim peak flowrates are the peak flows that can develop before the design life of the station is reached. Because the design life of the station can be 30 years or more, it is reasonable to install pumping equipment sized to pump the peak flow expected in the first 10 to 15 years of the station.

When the interim peak flow is reached, the station must be expanded to discharge the design peak flow. The system capacity can be increased by installing additional pumps or increasing the capacity of existing pumps.

Pump capacity can be increased by changing pump speed or increasing the diameter of the impellers installed in the existing pumps. Changing pump speed will require new drives. Installing larger impellers will increase the required input power to the pump, which may require new drives, or the original drives may be oversized to allow for future power requirements.

The initial minimum flow may determine the pump capacity if variable-speed pumps are installed. The lead pump must be variable speed and operate within the pumps' AOR (Figure 4.10). At minimum flow, the pump will operate at minimum speed and flow in the AOR. At full speed, the pump must also operate within the pump operating range. This may limit the maximum capacity of the pump, especially if the system has a flat system head curve.

Force main size and flushing velocities must be considered when designing pumping stations with phased conditions. Phasing of piping may be an option to provide two smaller diameter force mains, with one being in use during the interim phase and then the other as additional flow requires. Although this can be an option, it comes with higher capital costs and energy requirements and, thus, should be evaluated.

9.1.2 System Head-Capacity Curve

The system head-capacity curve presents a graph of the system into which the pumps will operate. Chapter 5 presents details for the preparation of the system curve.

When the system head-capacity curve has been developed, the initial, interim, and design year minimum, average, and peak flowrates should be added to help visualize the system requirements. Also, the force main velocities should be included on the curve.

Figure 4.18 is typical of a system head-capacity curve and shows a system curve for a collection system with a design population of 41,000 people and a 4200-m (13 800-ft) long, 0.6-m-diameter (24-in.-diameter) force main. The interim design population is 32,000 people, and the system static head is 9.1 m (30 ft). The curve includes the friction loss for new pipe and the design friction loss for old pipe and shows the flowrates and

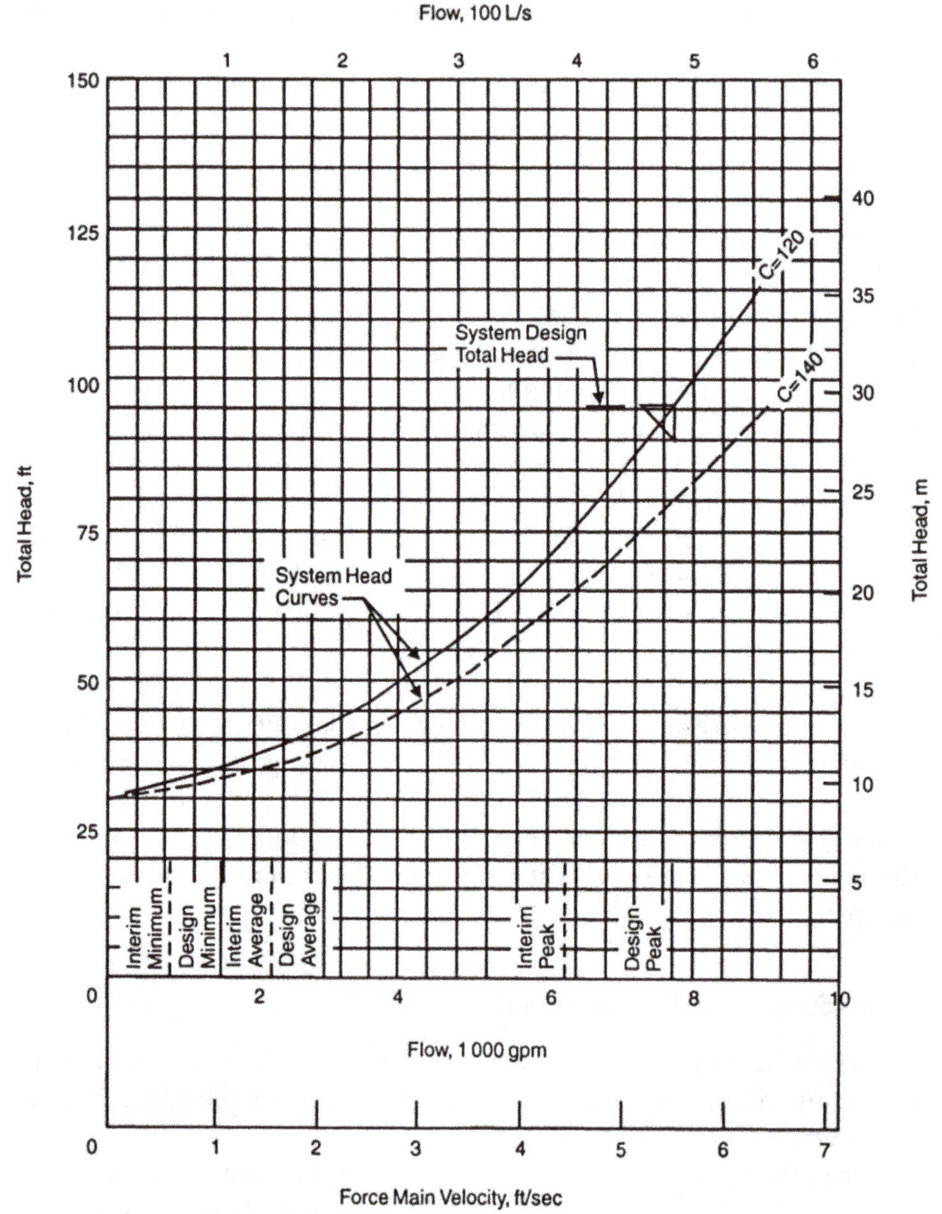

FIGURE 4.18 Typical System Head-Capacity Curve (ft/sec × 0.3048 = m/s)

force main velocities. The velocity in the 0.6-m (24-in.) force main varies from 0.17 m/s (0.6 ft/sec) at minimum flow to 1.7 m/s (5.5 ft/sec) at peak flow. If a 0.75-m (30-in.) force main were used to reduce friction loss, the velocities would be reduced to 0.1 m/s (0.33 ft/sec) at minimum flow and 1 m/s (3.3 ft/sec) at peak flow. These low velocities could cause problems with settlement of solids.

9.1.3 Station Location and Area Served

The station location and the area served may determine the type of pump control (on–off or variable capacity), the need for future growth, and the type of pump drive.

Stations that serve only a small portion of a collection system typically will be furnished with on–off pump control or two-speed pumps or controlled with variable-frequency drives. On–off pump controls are the simplest type. Pumps controlled by variable-frequency drives are more common than two-speed pumps when trying to control flow from the station.

If the station is located in an area that has potential for substantial future growth, the design peak flow and initial peak flow must be studied with care to ensure efficient operation. An oversized pumping system will use more energy than one designed for realistic flows. Consideration should be given to designing the station for interim peak flows, with the potential to increase station capacity in the future. However, this may affect force main sizing.

Pumping stations that discharge to a WRRF or large stations that discharge a large portion of the total collection system flow should be furnished with variable-speed pumps or step-speed control with small step increments. This will prevent sudden changes in facility flow that could upset the treatment process.

9.1.4 Force Main

Pump rates in the system must provide force main velocities that will both reduce settlement of solids in the force main and resuspend settled solids.

9.2 Selection Procedure

The recommended steps before selecting pumps for use in a pumping system include preparation and analysis of the system head-capacity curve, selection of the number of pumps, and preparation and analysis of system curves.

9.2.1 System Head-Capacity Curve Characteristics

The characteristics of the system head-capacity curve play an important part in the selection of the pumping equipment. The system curve is required to determine the rating of the pumps and is needed to determine the number of pumps in a system.

A system head-capacity curve with a high friction loss component and a small static head component favor the use of a single pump with two speeds or variable-speed drive, which would be the more common method to control speed. According to the affinity laws, the pump head decreases in proportion to the square of the speed ratio and the flow increases in proportion to the speed ratio. The system friction loss, however, is proportional to the flow reduction ratio squared. Therefore, if a single pump is selected to discharge the design flow at its BEP, a pump operating at reduced speed will also operate near its BEP.

A system with a large static component and a small friction component favors multiple pumps because if a pump is selected that, at full speed, operates near run-out on its operating range, a small reduction in speed may operate the pump at the minimum flow further from the BEP.

This is illustrated in Figure 4.19, which shows a high friction component system curve and a high static component curve superimposed on the

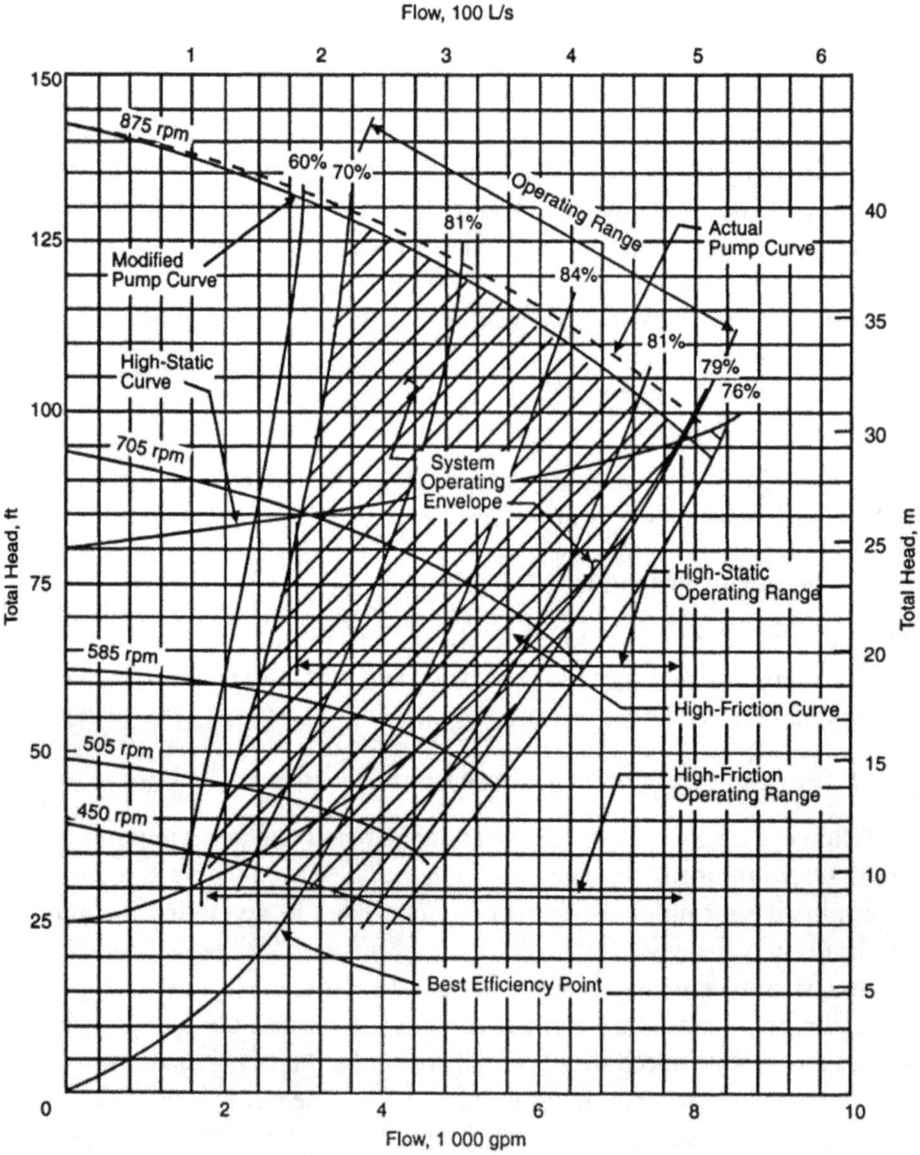

FIGURE 4.19 Effect of System Head-Capacity Curve on Pump Operation

variable-speed pump curve shown in Figure 4.20. The speed range and the flow range for the high friction component is larger than the system with a high-static component and the efficiency remains fairly constant. The efficiency of the system with the high-static component goes from one extreme to the other of the operating envelope.

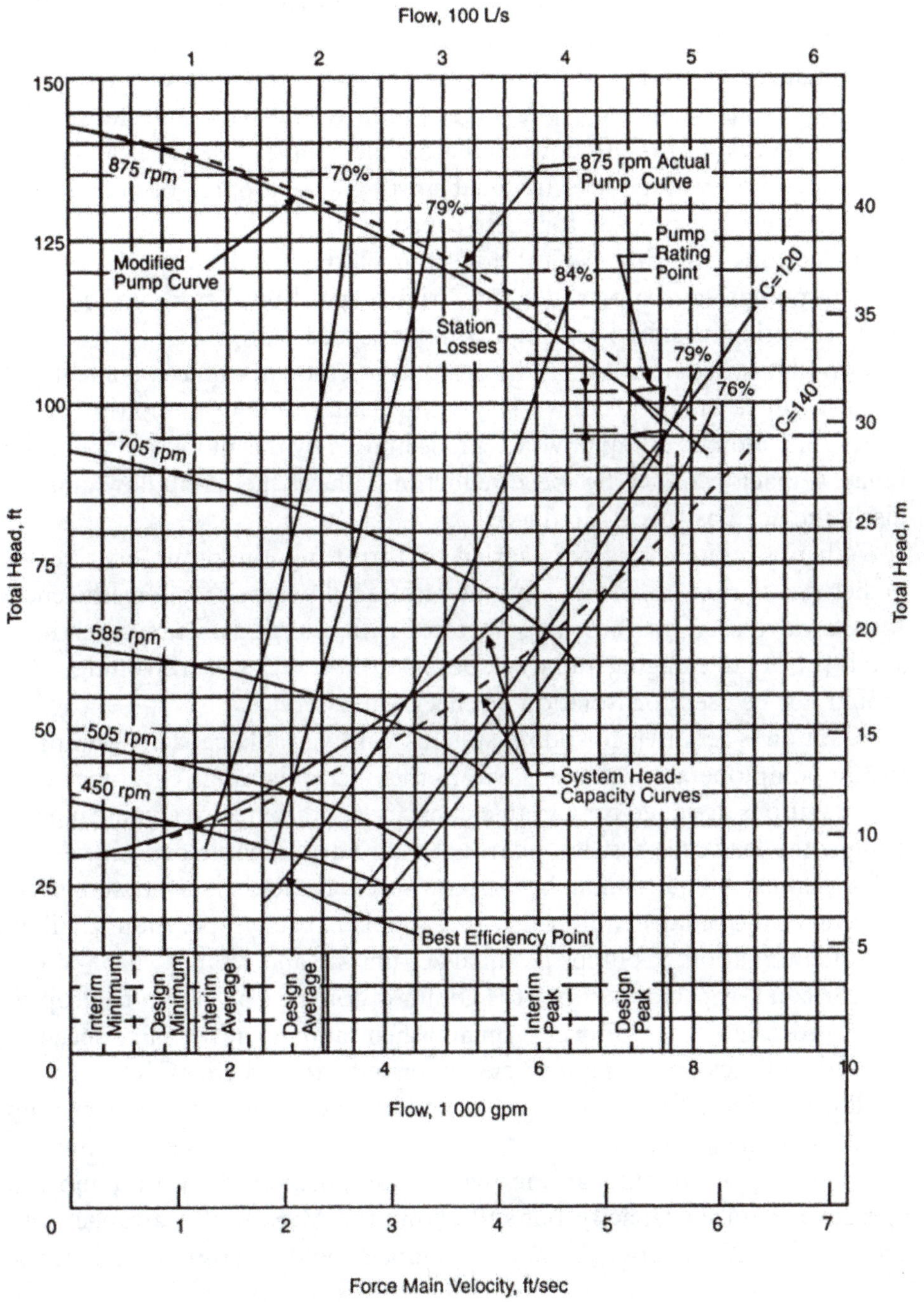

FIGURE 4.20 High Friction Head System Curve, One Pump (ft/sec × 0.3048 = m/s)

9.2.2 Preliminary Selection of Pumps

Preliminary pump selection can be made after the initial screening of station requirements. Because most municipal pumping stations require competitive bidding, pumps from several manufacturers should be included in the preliminary selection.

In stations with multiple pumps, the pumps can have equal or different capacities. System operation with different-capacity pumps is similar to multispeed operation of a pump system. When the inflow exceeds the capacity of the small pump, the larger pump would start, and the smaller pump would shut off. As flow increases, the smaller pump restarts. This operation, however, requires frequent starting and stopping of pumps, and the wet well must be sized for these capacity steps.

In stations requiring stepless flow control, the variable capacity of centrifugal pumps can be provided by variable-speed operation of the pump or by throttling the discharge of a constant-speed pump.

Variable-speed operation is the preferred method of capacity control and will provide a larger turndown in pumped capacity than pump throttling, which only reduces pump flowrate and is limited by the allowable operating range. Capacity control by speed reduction reduces the pump flow and the discharge head as speed is reduced.

When a pump capacity is varied by throttling, the pump operates at higher heads than necessary and, at reduced flows, with poor efficiency. Also, a valve used for throttling operates with a large pressure differential and is subject to clogging and wear. Because of the wear, the throttling valve should not be used for isolation or check valve service.

Caution is required for pump systems with variable-capacity pumps to ensure pump operation remains within the operating range envelope.

In stations designed for variable-capacity operation with equal-capacity pumps, the lead and first lag pumps should have variable capacity. Additional pumps, if required, can be constant speed; however, it is most common to have all the pumps equipped with variable-capacity operation to allow for full alternation of pumps to equalize runtime and wear.

For equal-capacity pumps, it is common for the lead pump to drop to low speed when a lag pump cuts on and then ramp up at the same speed to avoid the pumps operating inefficiently or outside of their AOR.

If the lead and first lag pumps have different capacities, the larger pump should be in the lead position and have variable capacity. When the smaller, constant-capacity pump starts in the first lag position, the lead pump can operate at reduced capacity, but still within the AOR. In this arrangement, coordinating BEPs between the two pumps should be considered because the pumps have different capacities and speed has a different effect on each pump's flow and head.

Alternately, it may be chosen to run a smaller pump, also known as a "jockey pump," in the lead position for energy savings and to save the runtime on a larger pump. In that scenario, if the smaller lead pump is not keeping up with a rising water surface elevation, the lead pump would shut off and the larger pump would cut on at a reduced speed and ramp up as required to maintain or lower the water surface elevation.

When operating with variable capacity pumps and fluids with solids, it is best practice to ramp the pump up to full speed to flush any solids from the pump and discharge piping before reducing to a lower speed resulting in lower velocities, especially with vertical discharge piping. A programmed operation to ramp the pump up to full speed again before shutoff to again remove any heavy solids from the pump and discharge piping may be considered.

In stations with variable-speed pumps, the standby pump should also be a variable-speed pump. The lead pump and first lag pump will do most of the work, with the second lag pump periodically operating. Therefore, the standby pump should be variable speed to provide reliable operation and share the station workload. In large stations requiring three or more operating pumps, variable-speed capability should be considered for most, or all, pumps.

9.2.3 Preparation of System Curves

Following the preliminary selection of pumps, prepare a system curve for each of the preliminary pump systems.

A system curve is the head-capacity curve for the system. The system curve and the pump curve are overlaid to verify the performance of the pump based on the system's characteristics. This system curve presents a picture of how the system will operate. A pump will discharge the flow where the pump curve intersects the system head-capacity curve. The system curve should be drawn only for the piping common to all pumps, so station losses should not be included in the system head-capacity curve.

Station losses are the friction losses in the suction and discharge piping of an individual pump and are deducted from the pump head-capacity curve to develop a "modified pump curve" for the individual pump. This is the curve the system curve will "see" the pump producing. The station losses must be added to the system head to determine the pump total head. Figure 4.20 illustrates a modified pump curve in which the station losses have been deducted from the actual pump head-capacity curve.

The reason for using the modified pump curve is to simplify the analysis. In a two-pump system, each pump discharges 50% of the total flow. If station losses were included in the system curve without recognizing that each pump handles only 50% of the flow, station losses would be calculated at 4 times the actual loss.

Modified curves can be used in developing the pump operating envelope. Once the full-speed modified curve has been developed, it can be used with the affinity laws to predict the pump curve at reduced speeds because the pump speed at reduced flow is proportional to the speed ratio and the head is proportional to the square of the speed ratio.

Correspondingly, station losses at reduced speed and, therefore, flow, are proportional to the square of the flow ratio and the square of the speed ratio.

9.2.4 Analyzing System Curves

Candidate system curves must be analyzed to allow selection of the optimum system. The following requirements should be considered:

- System $NPSH_A$ is greater than pump $NPSH_R$ under all operating conditions by an HI-defined margin and ratio;
- Pump operation is within the pump's POR or, at a minimum, within its AOR;
- Pump operation is efficient at interim and design average flows;
- Energy consumption is based on expected diurnal flow patterns;
- Wet well volume is adequate to prevent short cycling of constant-speed pumps and smooth operation of variable-speed pumps; and
- Force main velocities are within acceptable limits.

10.0 EXAMPLES OF PUMP SELECTION

The following section, which presents several types of system curves with pumps selected for operation with these curves, illustrates some of the correct applications of pumps to a system curve, and points out several potential misapplications.

10.1 High Friction Head System

The system head-capacity curve shown in Figure 4.18 is typical of a system with a small static head component and a large friction head component. The peak flow can be pumped by either one large pump or two smaller pumps operating in parallel.

Figure 4.20 shows a system curve with a single large pump and is based on the pump curve shown in Figure 4.6 and the operating envelope shown in Figure 4.10. The pump has been selected with the pump rating point slightly beyond its BEP, which places the BEP near the daily average flow when the pump operates at reduced speed.

Figure 4.21 shows the same system curve, but with two equally sized pumps to discharge the design peak flow. The pumps are selected with the rating point near the pump BEP to allow efficient operation at reduced speeds. An analysis of these two curves, with different modes of pump control, follows.

10.1.1 On–Off Operation of Single Speed Pumps

Single pump, Figure 4.20. Satisfactory operation. A better selection could be made if a pump could be found with the BEP closer to the rating point. A single pump, however, will require the largest wet well.

Two pumps, Figure 4.21. Unacceptable operation. The lead pump would operate outside of its design envelope, risking cavitation near the end of its head-capacity curve. A better choice of pump could alleviate this situation, as would a two-speed operation system (described below). This would allow operation closer to the envelope, preventing cavitation at the extreme.

10.1.2 Two-Speed Operation of Pumps

Single pump, Figure 4.20. Satisfactory operation at either 875 to 585 r/min or 875 to 705 r/min. At 585 r/min, the pump operates at its BEP and discharges more than the design average flow, but less than the interim or design peak flow. Therefore, this pump will operate most of the time at low speed with high efficiency to reduce power costs. High-speed operation will occur and keep the force main flushed out.

Pump operation at 705 r/min is also acceptable. The efficiency is reasonable, and the pump discharge is near the interim peak flow. However, the pump will operate at a higher head than for the 875- to 585-r/min selection, and the total power consumption for the 875- to 705-r/min pump will be greater than for the 875- to 585-r/min pump.

Two pumps, Figure 4.21. Satisfactory operation for 1175 to 875 r/min, but the controls must be set to eliminate operation of one pump at high speed. The control conditions would be lead pump at low speed, lead and lag pumps at low speed, and both pumps at high speed. The step of one pump at high speed plus one pump at low speed should be used with caution because the pump operating at low speed would be functioning outside of its operating envelope, near the shutoff head.

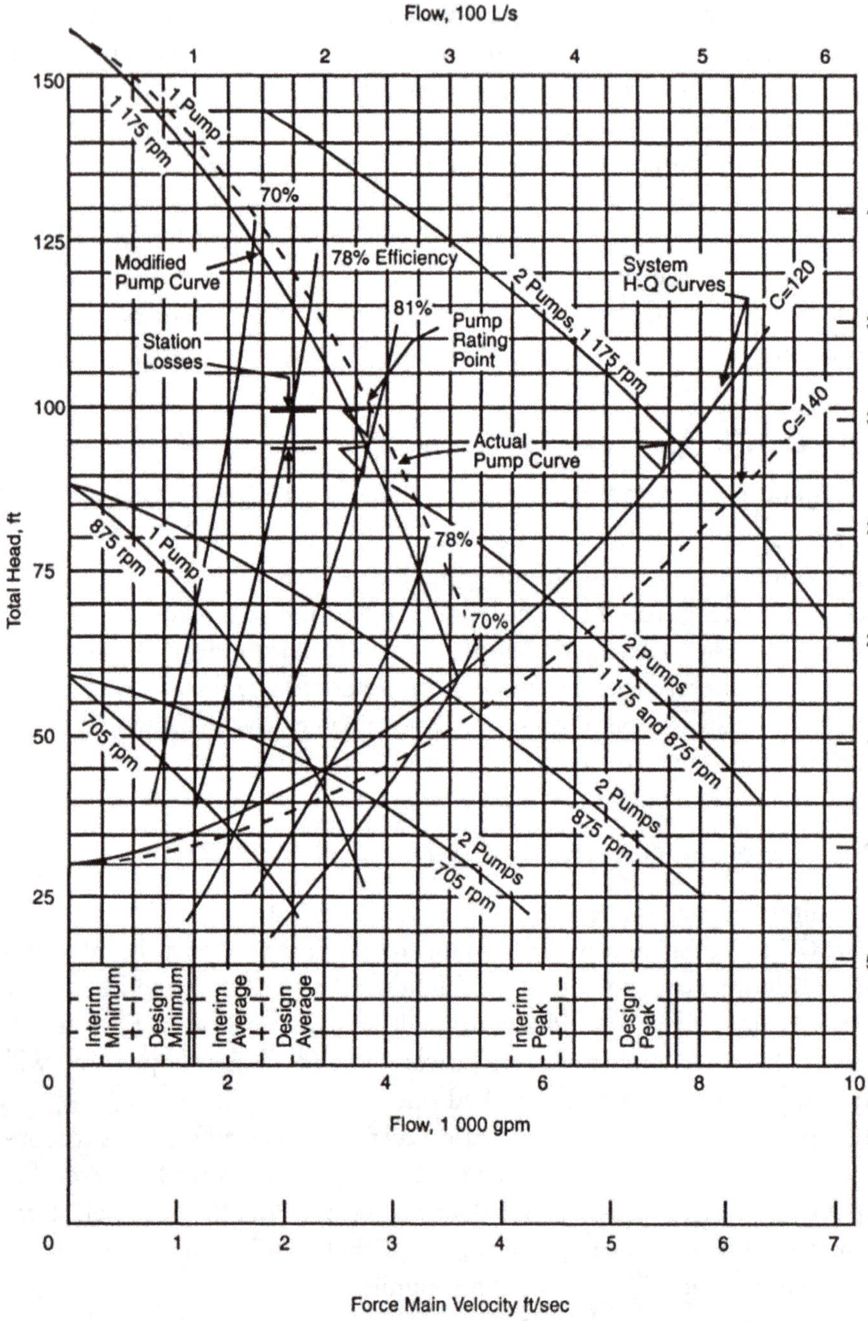

FIGURE 4.21 High Friction Head System Curve, Two Pumps (ft/sec × 0.3048 = m/s)

Pump speed of 1175 to 705 r/min should not be used because the change in flow between two pumps at low speed and two pumps at high speed is too great.

10.1.3 Variable-Capacity Operation of Pumps

Variable capacity by throttling is not acceptable for either the one- or two-pump system. The pumps can only be throttled to 225 L/s (3600 gpm) for the single pump and 150 L/s (2400 gpm) for the two-pump system without straying from their operating envelopes. These flows are at or greater than the system average flow. Also, throttling would require more energy than variable-speed pumps.

Variable capacity by variable-speed operation provides a wider flow variation than by throttling. The minimum flow by variable speed is 110 L/s (1750 gpm) for the single pump and 65 L/s (1000 gpm) for the two-pump system.

The single-pump system would operate outside of the operating range at minimum flows, but this may be acceptable because the system head is low, reducing bearing loads and potential erosion from recirculation; the pump would operate under these conditions for only short periods of time.

The two-pump system would operate adequately, but, to prevent cavitation in the lead pump, the controls must be set to start the second pump before the lead pump reaches full speed.

10.1.4 Power Consumption

Table 4.4 summarizes the power consumption for the single-pump and two-pump systems. The time required for pumping at various speeds is based on an assumed diurnal flow pattern for a population of 36,000, which is between the initial year and design year flows.

Considerable power can be saved if two-speed pumps are used because most of the flow is pumped at lower heads. The single pump operating at 585 to 875 r/min draws only 40 000 W (950 kWh/d), and the two-pump, two-speed motor operating at 875 to 1175 r/min draws only 37 500 W (900 kWh/d). However, the two-pump system would require a larger building than the single pump.

More power is required for variable-speed operation than for the most efficient two-speed operation because of losses in the variable-speed drive and constant operation of the pumps.

TABLE 4.4 Daily Power Consumption—High Friction System

	Single Pump		Two Pumps	
System Operation	h/d	kWh/d[a]	h/d	kWh/d
Single speed				
(On–off)	7.8	1550[b]	[c]	
Two speed				
(705/875 r/min)	9.5/0.6	1085	—	
(585/875 r/min)	10.4/2.1	950	—	
Two speed				
(875/2 at 875/2 at 1175 r/min)	—	—	11.1/2.6/ 1.4	900
(705/2 at 705/2 at 1175 r/min)	—	—	12.3/3.3/ 3.0	950
Variable speed	24	1040	24	920

Note. Power consumption based on an assumed diurnal flow variation for a system with a population of 36,000

[a]kWh/d × 41.67 = W

[b]Power consumption for a different single-speed pump operating at 85% efficiency at the rating point is 1440 kWh/d.

[c]On–off operation for two pumps at full speed is not recommended because lead pump operates beyond the pump operating envelope. The maximum power draw of 65 000 W (1550 kWh/d) is required by a single, constant-speed pump because all flow must be pumped against the maximum system head.

10.2 High Static Head System

Figures 4.22, 4.23, 4.24, and 4.25 show systems with high static head components and small friction head components. These curves illustrate the difference between pump selection for a high friction system and a high static head system. The flowrates for these curves are the same as those used for the high friction curve systems in Figures 4.20 and 4.21 and illustrate the requirements for one-, two-, and three-pump systems.

The pump in Figure 4.22 for single-pump operation is the same one used for single-pump operation as a high friction component system. This is a suitable selection for two-speed operation; but, for single-speed operation, a slightly larger pump is available with its best efficiency at the rating point. This larger pump cannot be used with two-speed operation because it would operate close to shutoff at low speed.

The pumps used in the two-pump system shown in Figure 4.23 are the same ones used in Figure 4.21 for the high friction head system. The BEP is at the rating point, and the pump operates efficiently when only one pump is in use.

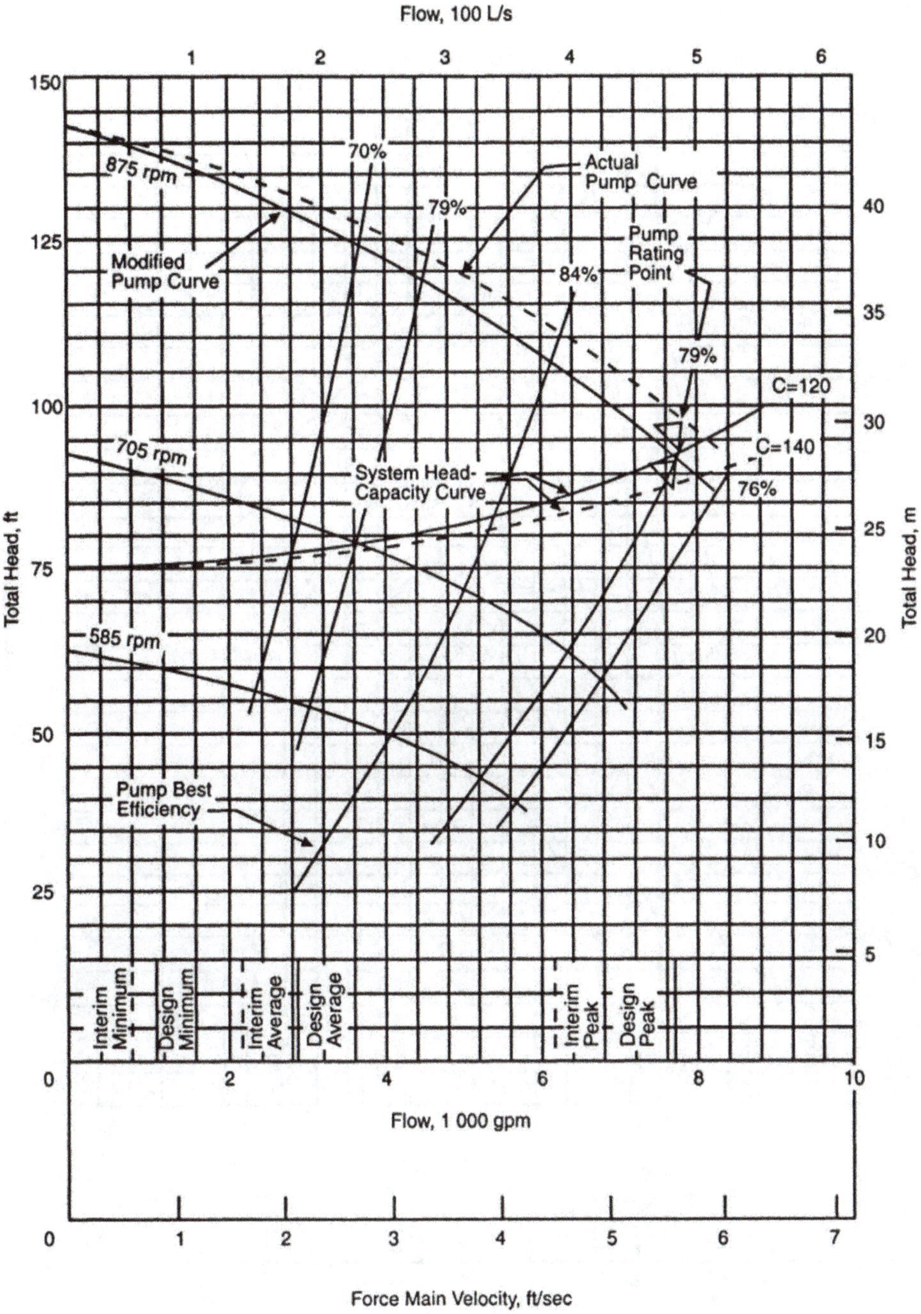

FIGURE 4.22 High Static Head System Curve, One Pump (ft/sec × 0.3048 = m/s)

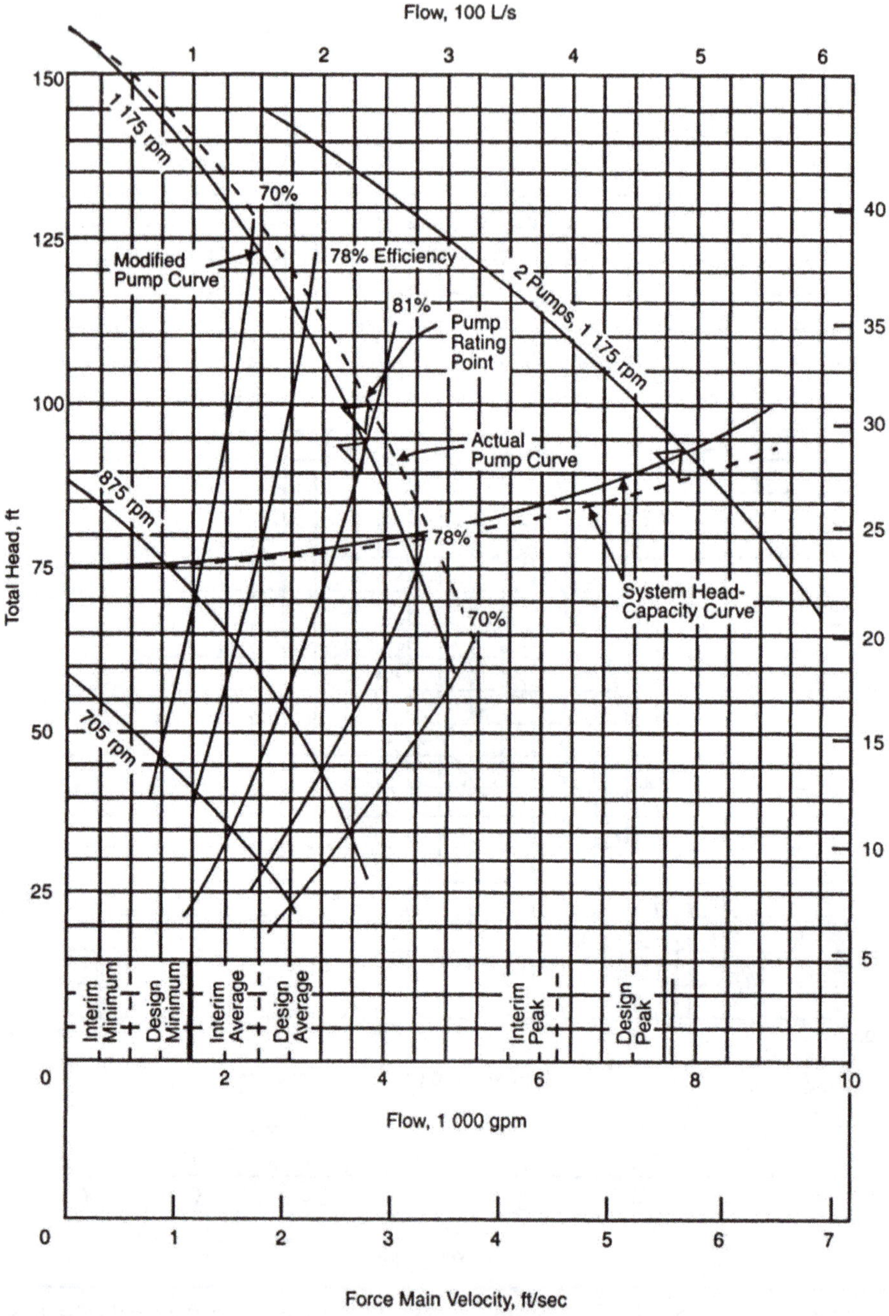

FIGURE 4.23 High Static Head System Curve, Two Pumps (ft/sec × 0.3048 = m/s)

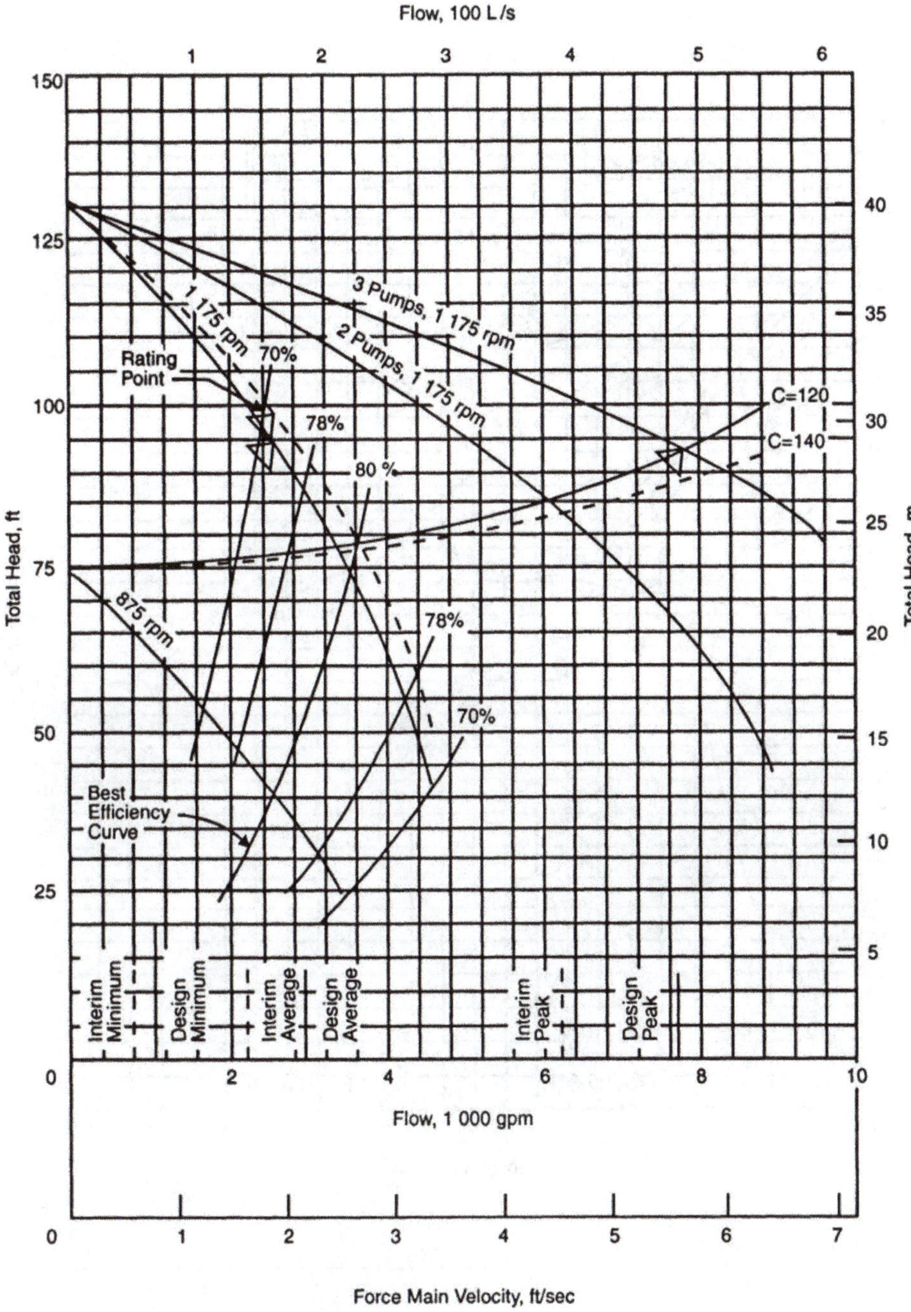

FIGURE 4.24 High Static Head System Curve, Three Pumps, On–Off Operation (ft/sec × 0.3048 = m/s)

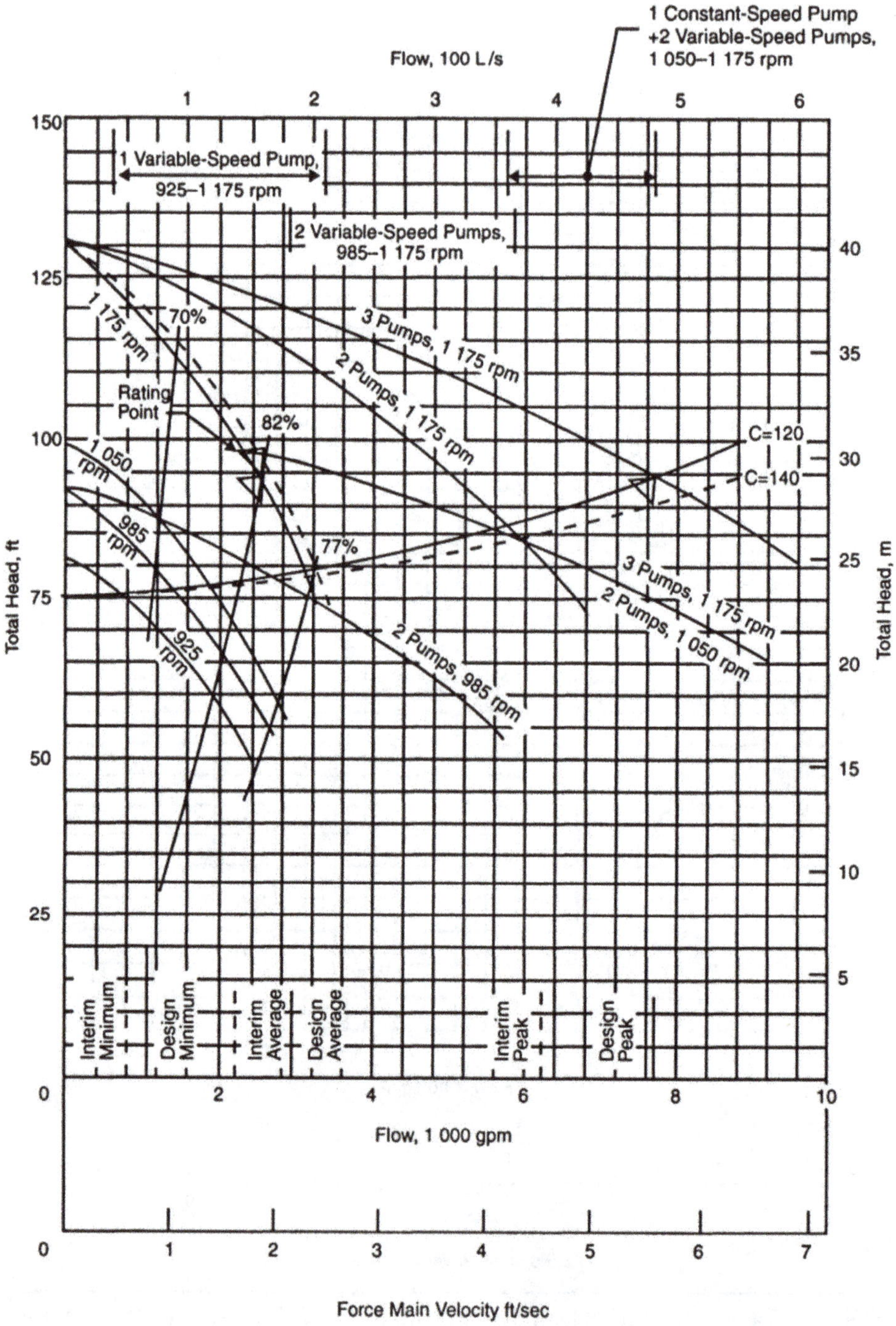

FIGURE 4.25 High Static Head System Curve, Three Pumps, Variable-Speed Operation (ft/sec × 0.3048 = m/s)

Different pumps should be selected for a three-pump system: one for on–off operation and a different selection for a system with variable-speed operation.

For on–off operation of a three-pump system, pumps should have a BEP beyond the rating point and, if possible, near the lead-pump operation point. Because the lead pump discharges the design average flow, the system is highly efficient. When two pumps are required, the system efficiency would still be acceptable. The least efficient system is with three pumps operating to discharge the design peak flow, but this occurs only for short periods of time. Figure 4.24 illustrates three pumps selected for on–off operation.

The pumps selected for variable-speed operation in a three-pump system should have their best efficiency near the pump rating point and, at full speed, the run-out end of the pump-operating envelope must be beyond the intersection with the head-capacity curve. This allows the pump to operate over the widest variable-speed range, so that at minimum speed the lead pump would discharge the minimum flow. This is illustrated in Figure 4.25.

It should be noted in Figure 4.25 that when discharging the interim minimum flow, the pump operates slightly outside the pump-operating envelope at 50 L/s (800 gpm). This may be acceptable because the pump would operate at reduced speed and head and at this flow for only short periods of time.

An analysis of these four curves with different modes of pump control follows.

10.2.1 On–Off Operation of Constant Speed Pumps

Single pump, Figure 4.22. Satisfactory operation. However, the system operation is inefficient because all flow must be pumped at the system maximum head. Wet well storage would be at a maximum.

Two pumps, Figure 4.23. Satisfactory operation. System operation would be better than single-pump operation because there are two steps of pumping. This system would require less wet well storage.

Three pumps, Figure 4.24. This system provides three steps of pump rates and would provide the most efficient operation. Wet well size is at a minimum.

10.2.2 Two-Speed Operation of Pumps.

Single pump, Figure 4.22. Two-speed operation with 875- to 705-r/min speeds would provide efficient operation, discharging approximately 250 L/s (3800 gpm) at low speed. This operation is close to that of the two-pump, single-speed system in which one pump discharges 265 L/s (4200 gpm).

Two pumps, Figure 4.23. Two-speed operation of these pumps is not acceptable because the low-speed operation of the lead pump is outside of the operating envelope, and it would operate too close to shutoff.

Three pumps, Figure 4.24. Like the two-pump system, two-speed operation is unacceptable because the pump operates close to shutoff at low speed.

10.2.3 Variable-Speed Operation of Pumps

Single pump, Figure 4.22. Variable-speed operation of a single pump is not acceptable because the minimum flow in the operating envelope intersects the system curve at the design average flow. This pump cannot operate to the system minimum flow.

Two pumps, Figure 4.23. As with the single pump, variable-speed operation of two pumps is unacceptable because they cannot pump minimum flows successfully.

Three pumps, Figure 4.25. This system provides the best operation for variable-speed pumps. At minimum system flow, the pump operates slightly below the minimum flow in the operating envelope, but it would operate at these flows for only short periods of time. In this case, the minimum flow dictates the size of the pump for a variable-capacity pumping system.

10.2.4 Power Consumption

Table 4.5 summarizes the power consumptions for the alternate systems. The time required for pumping at the various flows is based on the same assumed diurnal flow pattern used for Figures 4.20 and 4.21.

Because all flows are pumped at maximum head, a maximum power draw of 65 000 W (1550 kWh/d) is required by a single, constant-speed pump.

TABLE 4.5 Daily Power Consumption—High Static System

System Operation	Single Pump, kWh/d[a]	Two Pumps, kWh/d	Three Pumps, kWh/d
Single speed (on–off)	1550[b]	1360	1350
Two-speed operation	1350		
Variable speed	1650[c]	1525[d]	1450

Note. Power consumption based on an assumed diurnal flow variation for a system with a population of 36 000.

[a]kWh/d × 41.67 = W

[b]The pump shown in Figure 4.22 draws 1550 kWh. If a larger pump with the BEP at the rating point were used, the pump would draw 1440 kWh. Two-speed pumps should not be used because low-speed operation requires pumps to operate at or near shutoff.

[c]Pump should be operated on–off at minimum speed for flows less than 165 L/s (2600 gpm) to keep pump operation within the pump operating envelope.

[d]Pump should be operated on–off at minimum speed for flows less than 115 L/s (1800 gpm) to keep pump operation within the pump operating envelope.

Power can be saved using a two-speed motor for the single pump or by using two or three constant-speed pumps. These systems operate so that most of the flow is pumped at less than maximum head and the power draw of the three alternates are approximately equal.

If variable-speed pumping is required, the power draw is slightly higher than on–off steps because of the power loss in the variable-frequency drives. The power draw in the three-pump system is less than the one- and two-pump systems because only one pump operates continually, with the second and third pumps operating for only short periods of time.

In summary, on–off control of a single pump is the simplest method of control but consumes the maximum power and requires the largest wet well storage volume.

Depending on the system curve, multiple pumps or two-speed operation of a single pump will reduce power consumption and required wet well storage, but the controls are more complicated than those of a single pump.

The use of variable-speed pumping to match inflow reduces required wet well storage, but has the most complicated control system and, depending on the system curve, does not reduce power consumption. Variable-speed pumping does, however, match inflow and, when pumping to a WRRF, reduces shock loads on the treatment process.

10.3 Large Pumping Station System

Figures 4.26 and 4.27 present the pumping system for a large pumping station in which the major portion of the system head is static. This system curve is based on a new pumping station for a large midwestern city that was built to activate a new relief interceptor.

The interceptor receives mostly sanitary wastewater; but, during wet weather periods, large quantities of stormwater can be diverted to this interceptor. The contributing population is constant so there is only moderate increase in dry weather flows between initial and design-year flows. There will be an increase in wet weather flows as additional stormwater flows are diverted to the interceptor.

The pumping station is a lift station because it receives flow from the interceptor, lifts the flow, and discharges to an open channel. Therefore, the system head is constant, with its only variation being the difference between high and low wet well level for the speed control band.

Figure 4.26 shows system dry- and wet-weather flows. To avoid confusion, only the modified pump curve is shown on this system curve.

The pumps are sized for seven pumps, including one standby, to discharge the design-year peak storm flow. Because the station discharges

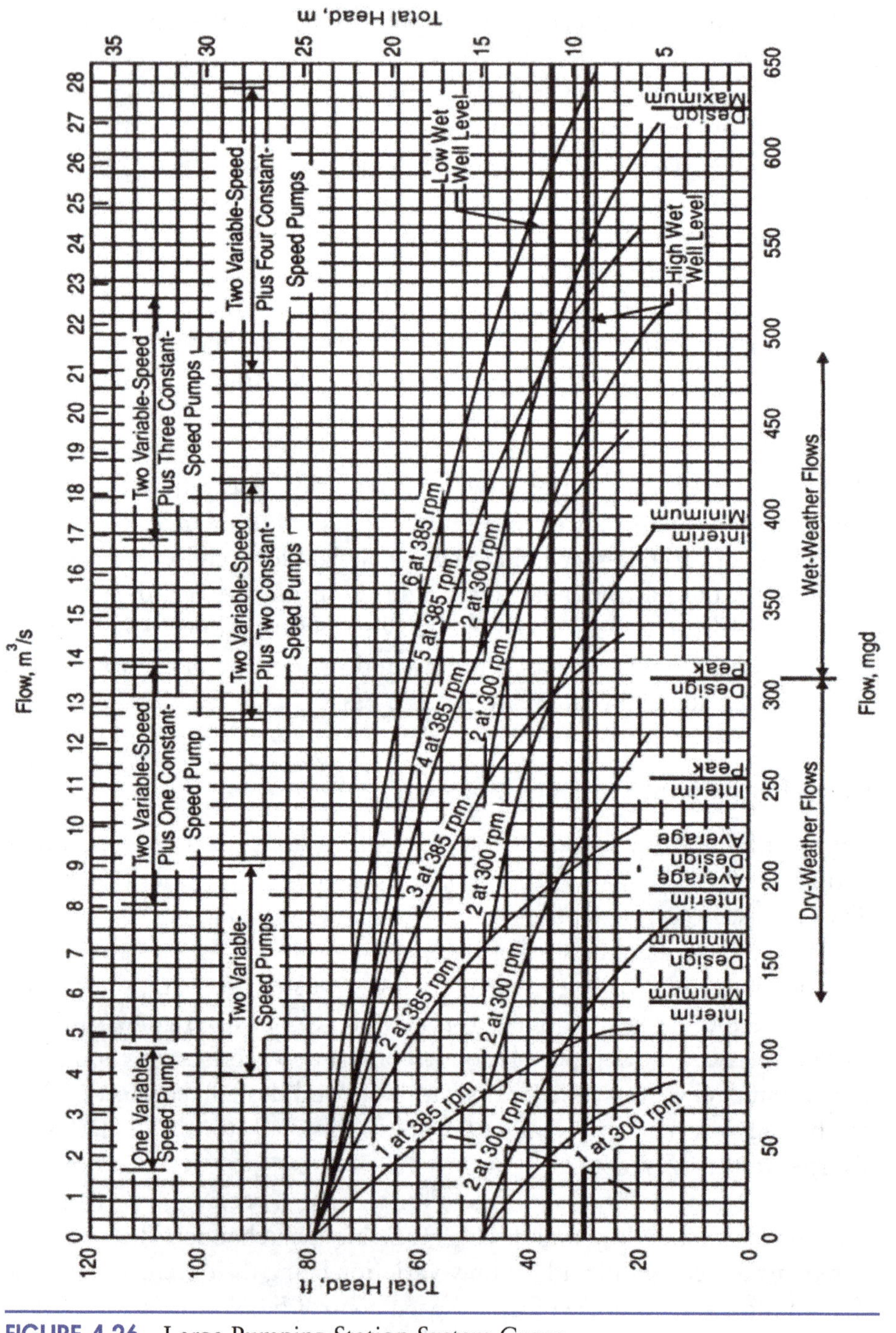

FIGURE 4.26 Large Pumping Station System Curve

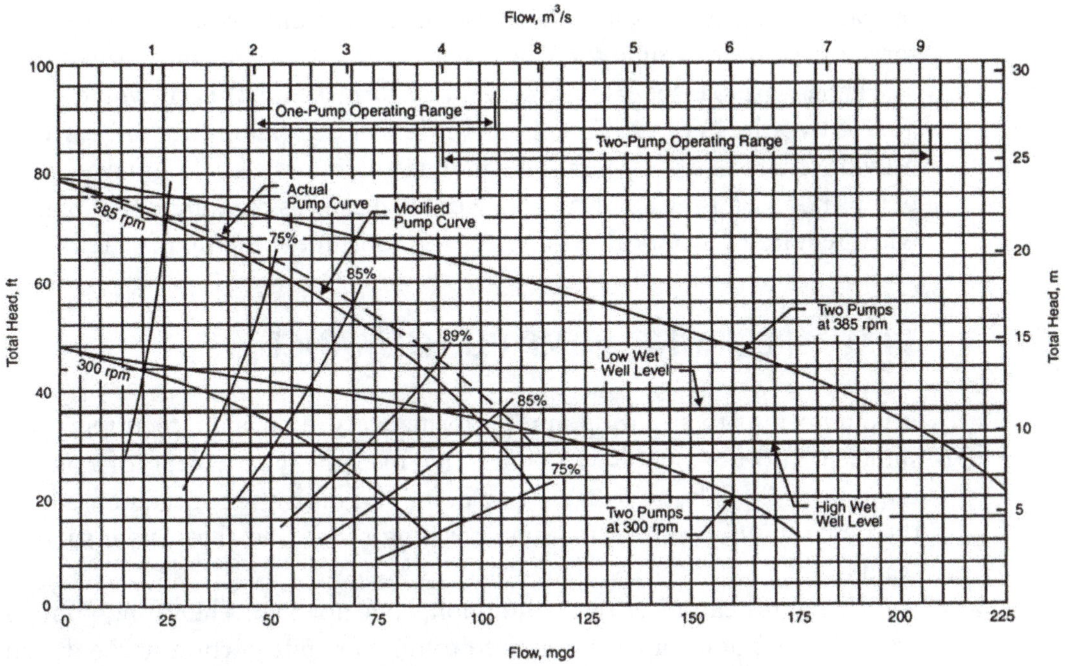

FIGURE 4.27 Large Pump Curve

to a WRRF, the pumping station has variable-speed pumps to match the influent flow.

The pumps are selected with the rated pump capacity beyond the pump BEP. Therefore, as the pump slows from full speed, the operating efficiency will increase until the best efficiency is reached and then decrease as the pump approaches the minimum flow in the pump operating envelope. Figure 4.27 shows the selected pump curve.

Because the interim design wet weather flow is only 17.2 m^3/s (395 mgd), five pumps will be installed initially, with space for two future pumps provided in the station.

Of the five pumps originally installed, four are furnished with variable-speed drives and one with a constant-speed motor. Future pumps will have constant-speed motors.

The reason for four variable-speed pumps is to allow operation without unnecessary wear to the pumps. The first two pumps online must be variable speed to allow the pump rate to match the inflow without one pump operating near shutoff head. In this station, the minimum flow requires the operation of two pumps at low speed. However, these two pumps, operating at full speed, can pump more than the dry weather design average flow. Therefore, if only two variable-speed pumps were provided, they would be operating continuously, and the constant-speed pumps would operate only for short periods of time for flows greater than design average flows.

Three variable-speed pumps could be furnished, but each would operate for two-thirds of the time. Four variable-speed pumps are furnished to provide reliable station operation with one out of service. The variable-speed pumps not used for lead or first lag would be used as constant-speed pumps when required for peak flows after all the constant-speed pumps are in use.

11.0 SPECIFYING PUMP PERFORMANCE

It is important to specify pump performance so that manufacturers have sufficient information to select pumps that meet the requirements of a pumping system. In addition to the rating point and maximum speed of a pump, the manufacturer must know its operating range and the system suction conditions.

All pumps can meet one rating point but not two. The rating point is typically the pump capacity required to discharge its portion of the design peak flow at the maximum system head, and pumps typically will operate at flows greater than the rating point. If the shape of the pump curve is critical, the engineer can specify alternate rating points that describe the allowable envelope, but should not specify an exact point. An acceptable method of specifying an alternate rating point would be to specify "not less than ___ L/s (___ gpm) or more than ___ L/s (___ gpm) at ___ m (___ ft) head."

To avoid cavitation, the manufacturer must also know the suction conditions. This information can be supplied by requiring successful operation to a specified minimum head or, in the case of a high-friction curve, successful operation to a curve between two alternate rating points.

The manufacturer must also know the NPSH available in the system. It is recommended that the $NPSH_A$ be understated by 1 or 2 ft (0.3 to 0.6 m). Hydraulic Institute Standards (1975) on testing allow a pump to exceed the rated flow by 10% at rated head or the head by 5% at rated flow. The additional friction losses at this plus tolerance may require the 1- or 2-ft (0.3- or 0.6-m) safety factor in the $NPSH_A$.

In variable-speed pumping systems, the manufacturer must know the expected speed range, which can be provided by specifying a reduced speed rating point. All manufacturers can meet this second pump rating point, but the speed between manufacturers will vary.

In complex systems, the designer may consider including a rating curve in the pump specifications. This curve should include the pump rating point, alternate rating points, minimum and maximum system head curves, and system $NPSH_A$.

It is recommended that the specifications require all pumps be tested by the manufacturer. These tests should include a hydrostatic test and an operating test. The operating tests on small pumps typically are simple tests that prove the pump discharges the rated capacity at the rated head. Larger pumps should be tested throughout the required operating range of the pump, with certified test curves submitted for review.

The need for witnesses to test the pumps should be considered by the designer, particularly for large pumps at critical installations. This requirement should be spelled out clearly in the specification as it would incur extra costs for the manufacturer, who typically tests these pumps twice—to prove pump operation before the witnessed test and during the actual witnessed test. Before the witnessed test, the manufacturer should submit test stand data, which must be reviewed to ensure that the test meets the requirements of Hydraulic Institute Test Codes and that the test stand can produce the required $NPSH_R$ to operate the pump over the specified range of operation.

12.0 REFERENCE

Karassik, I. J., Messina, J. P., Cooper, P., & Heald, C. C. (2007). *Pump handbook* (4th ed.). McGraw-Hill.

13.0 SUGGESTED READINGS

American National Standards Institute, Inc. (1998, March 3). *American national standard for centrifugal and vertical pumps for NPSH margin.* Hydraulic Institute.

Hydraulic Institute. (1975). *Hydraulic Institute standards for centrifugal, rotary and reciprocating pumps* (14th ed.).

Jones, G. M., & Sanks, R. L. (2008). *Pumping station design* (3rd ed.). Butterworth-Heinemann.

Metcalf & Eddy. (2014). *Wastewater engineering: Treatment and resource recovery* (5th ed.). McGraw-Hill.

Water Environment Federation. (2023). *Standards for design of water resource recovery facilities* (6th ed.; WEF Manual of Practice No. 8).

5

Piping Systems

The pumping station piping system discussion will follow the pool-to-pool concept, in which one analyzes the piping system between locations (pools) where the pumped liquid is affected by atmospheric pressure on the suction and discharge ends of the station being analyzed. There may be more than one pool on the discharge piping end as in multiple-pump stations or discharge points.

1.0 GENERAL DESIGN CRITERIA

The pumping station piping design must allow for wastewater or stormwater inflow variations. The flow will vary throughout the day, week, and season, and the station must be capable of handling peak flows with the largest pump out of service. The original design should consider future demands on the station, by installing connections for future pump suction lines. Discharge manifolds for larger pumps can be designed by providing for the replacement of pipe and by reducing fittings.

Pumping stations often serve far beyond their initial design life, and site selection should permit expansion, including wet well size and inlet piping for dry-pit stations. Sizing for the ultimate capacity of the station should be based on accepted standards (Great Lakes—Upper Mississippi River Board of State Public Health and Environmental Managers, 1990) or governing regulations. To ensure the satisfactory operation of pumps without vortexing or air entrainment in the liquid, pipes entering the wet well should be constructed considering the ultimate capacity of the station.

1.1 Piping

Piping is sized so that the velocity in the suction line will vary from 0.6 to 2.5 m/s (2 to 8 ft/sec), with a preferred value of 1.5 m/s (5 ft/sec) and will vary in the discharge piping between 1 and 2.5 m/s (3 and 8 ft/sec). As a general rule, when the velocity head ($V^2/2g$) exceeds 0.32 m (1 ft) for wastewater and 0.8 m (2.6 ft) for stormwater, a larger pipe should be used. With the exception of small grinder and effluent pump installations, piping less than 75 mm (3 in.) in diameter should not be used for conveying solids-laden wastewater.

An exception to this rule is for stormwater, when a self-priming siphonic recovery system is used to reduce static head. For these pump discharges, the water must rapidly fill the discharge pipe and sweep air out of the submerged pipe outlet or an air release valve located at a high point in the piping. As a general rule, a velocity of 2.1 m/s (7 ft/sec) is required, although lower speeds can be used depending on piping arrangements and can also be verified with physical or computer modeling. After the siphon is reached, the static head will reduce and can result in a higher water velocity than mentioned above.

Suction and discharge pipes should be supported and properly anchored and restrained. The discharge piping should be designed to facilitate dismantling and the allowance of thermal movements. Ductile iron-flanged pipe or flanged fittings, if encased in concrete, will complicate repair work. Ductile iron pipe typically is supplied with cement lining, though coal tar coatings and plastic or epoxy linings can be used. Reducers on the suction side of pumps should be eccentric and installed with the top flat to prevent air entrapment.

Welded steel piping is recommended for larger stormwater pumping stations as it is more economical at large diameters (DN 750, 0.9-m [36-in.] diameter) than ductile iron and considered more robust and less sensitive to thermal expansion than plastic-based piping. When working on a coastal environment with salt or brackish water, it is recommended to limit the amount of ferrous piping exposed to the sea or evaluate alternative pumps that either do not require discharge piping (flower-pot type) or use a concrete discharge conduit.

Both the suction and discharge piping must be supported and restrained to protect the pump casing. Settling of pipes may cause misalignment after installation, and care must be given to proper thrust blocking and anchorage of pipe where expansion or flexible joints are used and at pipe bends. Expansion or flexible joints must have piping restrained on both sides.

The capital cost of pipe and fittings in a large pumping station is a significant part of the total cost; thus, the choice of pipe size and materials is important. Larger pipes may have higher initial capital cost, but smaller pipes lead to increased hydraulic losses and energy costs.

1.2 Valves

Vacuum or air-release valves should be considered for installation at high points or long runs in discharge piping. Sizing of these valves (air flow rate) should consider the surge effect on the system. Air valves should be designed to meet the criteria of American Water Works Association's (AWWA; 2016) design manual M51, *Air Valves: Air-Release, Air/Vacuum and Combination.*

Drain or blowoff valves at all low points would help in removing sedimentation and allow for maintenance of the pipeline. Blowoff valves should be sized so main pipeline velocities drain no greater than 1 ft/sec. Blowoff valves should be strategically located for access and environmental considerations and sloped to prevent back-siphoning into the line.

The valve on the suction line should be of high quality to resist corrosion and still function when closed. A sluice gate installed at the wet well entrance and between sides of a divided wet well facilitates dewatering for cleaning, inspection, and repair.

Stormwater piping that goes over a line of protection (levee or floodwall) is required to have a vacuum breaker valving system that includes a redundant, manual means to break vacuum and thus prevent siphonic backflow of water into the protected area. Piping that goes through the line of protection requires two means of backflow prevention, with one being a passive valve such as a check or flap valve. Best practices for the second valve are an isolation valve such as a butterfly or knife gate valve. Consideration should be given to the maintenance of these valves because debris that can pass through a trash rake and pump can get caught on the stem of a valve.

Vertical pumps typically used in stormwater do not require any suction piping and, due to the size of the intake basins, sluice gates are often impractical. In these instances, stop log slots should be provided for cleaning, inspection, and repair.

1.3 Maintenance

To allow for proper maintenance of the pump and check valve (or pump control valve), a ball or plug valve for isolation should be provided in the discharge line of each pump downstream of the check valve. Check valves should be installed in a horizontal line as opposed to a vertical line.

1.4 Force Mains

Though force mains generally are designed to reduce annual operating costs, the designer also should consider the possible accumulation of deposited material in the line. Optimum velocities for reducing cost and preventing accumulation of solids range between 0.9 and 1.5 m/s (3 and 5 ft/sec), which are high enough to resuspend material that may settle out when the pumps are not operating. Other options for preventing accumulation of solids in a force main include a pigging facility. A pigging facility is a piping configuration on both ends of the pipeline that allows a polyurethane or foam bullet-pig, with roughly the same inside diameter of the pipe to be inserted into the force main to sweep out/remove the sedimentation that can build up in the pipeline without having to shut down pumping operations. A pigging facility consists of a launching station at the discharge of the pumping station and a receiving station at the end of the force main. The pig can be inserted into the force main without shutting down pipeline operations and uses differential pressure from the launching station to the receiving station to move and sweep through the force main pipeline. How the sediment and liquid will be captured at the receiving end must be considered. Pigging facilities can also be used to insert chemicals or gel pigs to help clean the pipe or alleviate hydrogen sulfide (H_2S).

The increasing popularity of variable-speed pumping must be considered in relation to force main velocity, and provisions for cleaning or flushing must be made for extended low-flow time periods with velocities less than 0.6 m/s (2 ft/sec). One suggested method is manual operation at higher speeds and at a predetermined time interval.

Materials used for force mains include ductile iron, steel, high-density polyethylene (HDPE), polyvinyl chloride (PVC), fiberglass or reinforced plastics, and prestressed and reinforced concrete. System design pressures, installation methods, and surge allowance may preclude the use of some materials.

Flow typically is intermittent in wastewater force mains, with gases (predominantly H_2S) generated during periods of nonflow. When combined with water, a mild sulfuric acid is formed that is corrosive to many pipe materials.

In longer force mains, especially at higher ambient temperatures, wastewater generates H_2S gas from anaerobic decomposition. Controlled injection of air or chemicals to the force main at the pumps in larger pumping stations can reduce hydrogen sulfide buildup. Care must be taken, however, in selecting the point of entry and the rate of injection. Some alternatives to air injection or chemical feed are hydrogen peroxide and ozone or the use of noncorrosive force main piping materials (Water Environment Federation, 2020). Installation of an isolation valve on the force main just outside the pumping station would allow the valve installed on the pump discharge to be inspected and serviced, assuming the discharge line can be taken out of service. A bypass connection outside the pumping station would permit bypassing the station with a portable pump.

Good practice dictates pressure testing the force main, which is completed in advance of operations. Before testing, the pipeline must be backfilled or braced to prevent movement under pressure. Pipeline bends must be fitted with permanent thrust blocks or harnessed restrained joints, constructed to withstand test pressure thrusts. When testing a completed pipeline, hydrostatic testing is preferred over air testing for safety reasons. Elevation differences that can create test pressures that would exceed maximum allowable pipeline pressures along the force main route should be taken into consideration in terms of whether the force main will need to be tested in sections or can be done in one test. It is more efficient and cost effective to test the entire force main all in one test.

The force main should be filled slowly from any available source, with water introduced from temporary connections made at the lowest point in the line. Flow velocity during line filling should not exceed 0.6 m/s (2 ft/sec), depending on the pressure rating of the system.

Air and vacuum valves should be designed for wastewater service with cleanout provisions to facilitate maintenance because they are likely to clog with grease from force mains carrying raw wastewater. On low spots where gritty material is likely to accumulate and restrict flow through the force main, a blowoff should be installed. Points of discharge should be well ventilated. To reduce corrosion, force mains should be kept filled during periods of normal intermittent operation and dewatered only for maintenance or repair. Air vent and relief valves should be protected from freezing and from damage by heavy equipment, such as is used for construction and grass cutting.

2.0 FLUID FLOW FUNDAMENTALS

Pumping stations are used where it is not possible or cost effective to construct a gravity line to remove stormwater or wastewater. Often called *lift stations*, pumping stations raise wastewater from one gravity line to another. Figure 5.1 illustrates the schematic factors involved in assessing the hydraulics of a stormwater or wastewater pumping station. Typically, the liquid is to be transported from one location to another at a higher elevation. A pump is needed to produce the required total head at the required flowrate. The total head is the sum of the pipe friction losses and the elevation difference from the inlet water level to the highest elevation in the discharge force main or static head.

The energy level of the liquid from an elevation datum and losses in the system are shown on the hydraulic grade line (HGL) in Figure 5.1. The

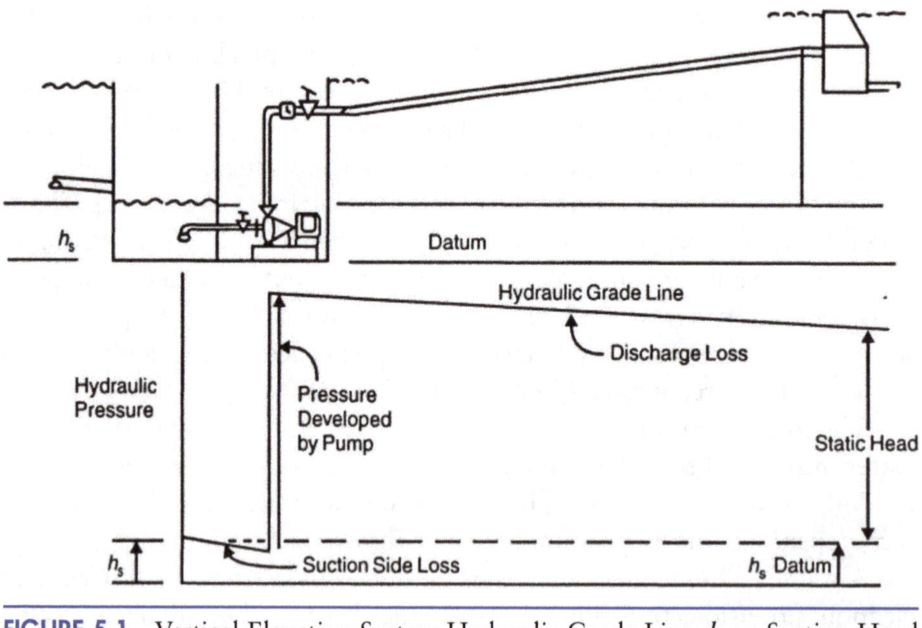

FIGURE 5.1 Vertical Elevation System Hydraulic Grade Line: h_s = Suction Head

energy grade line reflects a certain system condition such as a single flowrate. When designing a pumping station, a designer should establish a range of flows and select a specific pipe size and station arrangement. The resistance to flow is represented by the head-flow curve resulting from this analysis and is called a *system resistance* or *system head curve*.

2.1 System Curve

The system curve is a picture of the total head (pressure) a pump must supply when operating to produce a certain flowrate. The total head the pump must develop is the sum of the static head and friction losses. The pump can only operate at the intersection of the system head curve and pump performance curve.

Pipe friction losses for pressure pipe applications can be determined by the selection of roughness coefficients such as Hazen–Williams or Darcy–Weisbach, which then are used in the appropriate equations. Roughness coefficients typically are based on pipe roughness after use for several years and can be a determining factor economically. Pump selection, however, should consider the difference in suction pressure and the pipe roughness for both new and old pipe. The friction head should be calculated as accurately as possible. Fitting or minor losses are calculated using applicable coefficients to apply to the velocity head for the type of fitting or valve being considered.

The diameter of a force main is selected based on the size that will give the lowest overall annual cost or capitalized cost, depending on the method of calculation. The procedure involves selecting various pipeline diameters and calculating the pump head based on the velocity of pumping at the normal rate of flow (not the peak rate). The pump is sized from its head, and the pumping power is determined from the pump curve at its corresponding head and efficiency for the normal flowrate. The pumping power cost then is calculated using both the energy per kilowatt hour and the demand charge per kilowatt. This cost may be capitalized and added to the capital cost for the pump, piping, and force main.

Alternately, the capitalized cost of the latter three items may be calculated on an annual basis and added to the annual power cost. The pipe diameter that gives the lowest cost, on either basis, is the most economical diameter provided the velocity of flow is suitable for wastewater pumping.

Pressure surges are any deviation from the normal steady-state hydrostatic pressure in a pumping system, and they occur during partial or complete interruption of flow for reasons including pump starting and stopping, valve closures, power failures, valve openings, or if pipes are broken. Most stormwater and wastewater force mains are under low hydrostatic pressures, but the flows have relatively high operating velocities and momentum that contribute to high surge pressures.

Care should be taken in the design of pipe to determine wall thickness based on internal operating pressure, surge pressure, and total (passive and active) earth or equipment loads on the pipe. Also, proper bedding for the pipe must be provided. Joint design for each type of pipe should be investigated for proper compression of gaskets, expansion and contraction of joints, and the thrust anchors. The weak links in many force mains are at bends where thrust restraints are needed, and the maximum pressure the pipeline will be subjected to should be determined carefully.

2.2 Equations

Liquid flow relationships in piping systems are described by three mathematical relationships:

- The continuity equation;
- The general energy equation (Bernoulli); and
- The momentum equation.

While these relationships are based on the assumptions of incompressibility of the pumped fluid and steady-state or one-dimensional (stream tube) flow, equations are commonly applied to all conditions of pipe flow. Surge analysis requires special consideration. Under steady-state (i.e., constant at any one point in time) flow conditions, the continuity equation states that the product of flow area and velocity is constant for a given flowrate. The general energy equation is credited to Bernoulli and applies to the energy level of a fluid at any particular location in a system. If the liquid is incompressible and no heat is added or taken away and no work is done by or on the liquid, then the appropriate energy terms are *velocity head*, *pressure head*, and *position* or *elevation head*.

In real systems, between two points there would be an energy change so that comparing the terms between two points will give a differential energy term, *head loss*.

Finally, the momentum equation relates the change in velocity vectors (direction or magnitude change) to a mass rate of flow and the resulting components of force that must be restrained. A pipe bend is an example.

The continuity equation is as follows:

$$Q = AV \tag{5.1}$$

(For liquids)

$$Q_1 = Q_2 \tag{5.2}$$

Then

$$A_1V_1 = A_2V_2 \tag{5.3}$$

Where

Q = flow rate, m^3/s (cu ft/sec);
A = area, m^2 (sq ft), and
V = velocity, m/s (ft/sec).

The general energy equation is as follows:

$$Z_1 + \frac{V_1^{\ 2}}{2g} + \frac{P_1}{Y} = Z_2 + \frac{V_2^{\ 3}}{2g} + \frac{P_3}{Y} + h_L \tag{5.4}$$

Where

Z = elevation above datum, m (ft);
$V^2/2g$ = velocity head, m (ft);
g = gravitational constant, 9.8 m/s^2 (32.2 ft/sec^2);
P = liquid absolute pressure, kN/m^2 (lb/sq ft);
Y = specific weight, kg/m^3 (lb/cu ft); and
h_L = energy loss, m (ft).

2.3 Pipe Flow Parameters

To properly size pumps for the station, the designer needs to calculate the pressure required to move the liquid through the pipe at the desired flowrates. Various investigators have presented equations for calculating friction losses in closed conduits. The resistance to flow in a pipe of any incompressible liquid is proportional to pipe size and velocity head. The friction factor is the constant of proportionality and is itself a function of pipe roughness.

The Darcy–Weisbach and Hazen–Williams formulas are used for pressure pipe applications, and the Manning equation is used for open channel and gravity flow in pipelines with a free water surface. These equations are in a form for open-channel and full-pipe flows. The formulas have been incorporated into tables for pipe size and type in many hydraulic handbooks (Crane, 1974; Heald, 1992; Hydraulic Institute, 1979; Submersible Wastewater Pump Association, 1986).

Darcy–Weisbach:

$$H = f[L/D]V^2/2g \tag{5.5}$$

H = friction loss, m (ft);
f = friction factor;
L = pipe length, m (ft);
D = pipe diameter, m (ft);
V = velocity, m/s (ft/sec); and
g = gravitational acceleration 9.81 m/s^2 (32.17 ft/s^2).

Hazen–Williams:

$$h = L[V/0.85C \cdot R^{0.63}]^{1.85} \text{ metric} \tag{5.6}$$

$$h = L[V/1.318C \cdot R^{0.63}]^{1.85} \text{ English units} \tag{5.7}$$

C = pipe roughness coefficient and
R = hydraulic radius, m (ft) ($D/4$ for circular pipes).

Manning (open channel):

$$h = L[Vn/R^{0.67}]^2 \text{ metric} \tag{5.8}$$

$$h = L[Vn/1.49R^{0.67}]^2 \text{ English units} \tag{5.9}$$

n = roughness coefficient and
R = hydraulic radius (cross-sectional area wetted perimeter, m [ft]).

Most tables of friction losses are calculated on the basis of new pipe. The designer must estimate and incorporate into the design the increase in pipe roughness from aging. The Hazen–Williams equation C factor (or aging coefficient) is a ready way to estimate increased pipe roughness for use in that formula. Table 5.1 shows C values typically used for design purposes. Note that new pipe values vary from the design C values.

The absolute roughness parameter varies with pipe type and surface finish. Table 5.2 provides the values of the roughness coefficient for different pipe types for use in the Darcy–Weisbach and Manning equations. The relationship between pipe roughness and friction factor can be obtained from the Moody diagram (see Figure 5.2).

2.4 Pipe-Fitting Parameters

In most systems, frictional losses from pipe fittings typically represent only 10% or less of the total system friction losses. Though small, they must be calculated and added to the system. In stormwater pumping stations, minor losses can be a high percentage of total friction losses. Minor head losses (fitting losses) are proportional to the velocity head ($V^2/2g$). The proportionality

TABLE 5.1 Hazen–Williams Friction Coefficients (*C* for Condition of Pipe)

	C		
Pipe Type	**New**	**Design**	**17–20 years**[a]
Ductile iron (lined)	150	140	120
Fabricated steel	140	130	100
Fabricated steel (lined)	145	135	120
Plastic—PVC	150	140	120
Concrete cylinder pipe	145	140	120

[a]Values can vary widely because of site-specific conditions such as velocity, length of line, and liquid characteristics.

TABLE 5.2 Values of Effective Absolute Roughness and Friction Formula Coefficients

Conduit Material	**Effective Absolute Roughness (Darcy–Weisbach) *k*, ft**	**Manning *n*, ft$^{0.5}$**
Closed conduits		
Asbestos-cement pipe	0.001–0.01	0.011–0.015
Brick	0.005–0.02	0.013–0.017
Cast iron pipe		
Uncoated (new)	0.000 85	—
Asphalt dipped (new)	0.0004	—
Cement-lined & seal-coated	0.001–0.01	0.011–0.015
Concrete (monolithic)		
Smooth forms	0.001–0.005	0.012–0.014
Rough forms	0.005–0.02	0.015–0.017
Concrete pipe	0.001–0.01	0.011–0.015
Corrugated-metal pipe (0.01 m [0.5-in.] × 0.07 m [2.6-in.] corrugations)[a]		
Plain	0.1–0.2	0.022–0.026
Paved invert	0.03–0.1	0.018–0.022
Spun asphalt-lined	0.001–0.01	0.011–0.015
Plastic pipe (smooth)	0.01	0.011–0.015

(*continued*)

TABLE 5.2 Values of Effective Absolute Roughness and Friction Formula Coefficients (*Continued*).

Conduit Material	Effective Absolute Roughness (Darcy–Weisbach) *k*, ft	Manning *n*, $ft^{0.5}$
Vitrified clay		
Pipes	0.001–0.01	0.011–0.015
Liner plates	0.005–0.01	0.013–0.017
Open channels		
Lined channels		
a. Asphalt	—	0.013–0.017
b. Brick	—	0.012–0.018
c. Concrete	0.001–0.03	0.011–0.020
d. Rubble or riprap	0.02	0.020–0.035
e. Vegetal	—	0.030–0.40
Excavated or dredged		
Earth, straight, uniform	0.01	0.020–0.030
Earth, winding, fairly uniform	—	0.025–0.040
Rock	—	0.030–0.045
Unmaintained	—	0.050–0.14
Natural channels (minor streams, top width at flood stage < 100 ft)	0.1–3.0	—
Fairly regular section	—	0.03–0.07
Irregular section with pools	—	0.04–0.10

[a]Values can vary widely because of site-specific conditions such as velocity, length of line, and liquid characteristics.

constant *K* has been experimentally and analytically determined for a large variety of fittings. The formula is

$$h = \frac{kV^2}{2g} \tag{5.10}$$

Where

h = minor head loss, m (ft);
K = proportionality constant;
V = pipe velocity, m (ft); and
g = gravitational constant.

Some designers convert minor losses to an equivalent number of feet of straight pipe. There are also tables for this method (Heald, 1992). Hydraulic

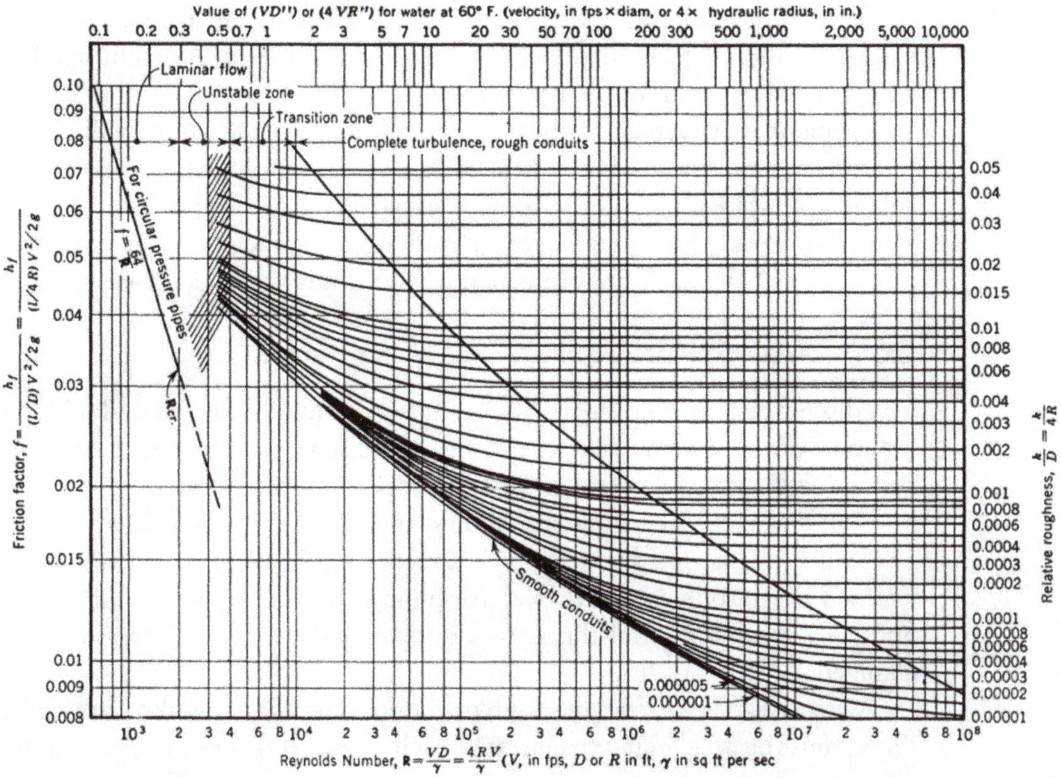

FIGURE 5.2 Moody Resistance Diagram for Uniform Flow in Pipes

Note. [0.555 6(°F–32) = °C; in. × 25.4 = mm; ft × 0.3048 = m; ft/sec × 0.3048 = m/s; sq ft/sec × (9.29 × 10^{-2}) = m^2/s]

handbooks (Crane, 1974; Heald, 1992; Hydraulic Institute, 1983) list the minor loss coefficients for common and more complicated shapes.

There are several computer programs commercially available for the solution of simple and complex piping system problems (CEComp, 1986; University of Kentucky, n.d.; Wood et al., 1991). Some of these programs are useful, reasonably priced, and can be used on personal computers. They greatly simplify the task of pipe sizing, pump head determination, and system performance prediction.

There are also computer programs for selection of centrifugal pumps, and some use pump manufacturer catalogs (e.g., Pump-Flo). Another is a generic program that specifies the operating parameters of a pump and produces a technically correct pump size, operating speed, and efficiency (Krebs, 1990).

2.5 Station Piping

Smaller pumping stations have two or three pumps, typically of the same size, and larger stations may have more. The discharge piping of the pump to the station discharge is termed *station piping*.

There are numerous methods of affecting discharge from a pumping station and some general rules that should be followed. As mentioned in Chapter 3, the pump discharge check valve should be mounted in a horizontal plane, not vertically where solids may accumulate on the valve disk. Flow to a common discharge header from pump piping should be to the side of the header, not the bottom, to avoid entrained solids from settling to the discharge pipe of an inactive pump. Entering the discharge header at an angle, as from a lateral fitting, can reduce minor fittings losses.

2.6 Discharge Line Piping

The discharge line from the pumping station should be considered an integral part of the system. Suction, station, and discharge piping determine the system head curve, which is critical to proper pump selection.

The discharge line diameter should provide a minimum line velocity of 0.6 m/s (2 ft/sec). Line velocity in excess of 2.5 m/s (8 ft/sec) should be avoided except on very short lines. All piping must be supported for weight and for lateral and axial thrust forces generated by internal pressure and changed direction of flow.

With constant-speed (intermittent pumping) design, the discharge pipe liquid must be accelerated from rest at each pump start. The designer should check for surge and acceleration head requirements for each design. Designs with a high proportion of static head to total head and those with longer force mains should be reviewed for the acceleration head effect. From Newton's second law, the following relationship can be developed for the elapsed time to accelerate the liquid in the discharge pipe from zero velocity to design velocity:

$$t = g^{-1}L \cdot V/(Hso\text{–}Hs) \qquad (5.11)$$

Where

t = pause time to accelerate, seconds;
g = gravitational constant, m/s^2 (ft/sec^2);
L = discharge line length, m (ft);
V = velocity, m/s (ft/sec);
Hso = centrifugal pump zero flow head, m (ft); and
Hs = system static head, m (ft).

In operation, the centrifugal pump starts and develops the shutoff or zero flow pressure until acceleration of the liquid in the line permits the pressure to drop to the normal intersection of the system head curve with the pump performance curve. The pause point in the startup sequence is at a flowrate less than the minimum noted by the pump manufacturer. Some manufacturers do not show minimum and maximum operating flowrates on their performance curves.

All volute casing centrifugal pumps develop high radial thrust forces as they operate close to their zero flow (shutoff head) condition (Hydraulic Institute, 1983). These high radial forces impose reverse bending loads on the shaft of overhung impeller pumps, which may produce stresses in excess of the endurance limit of the shaft material, resulting in early fatigue failure of the shaft. The loads are also transmitted to the pump bearings, causing them to fail. The designer should calculate the parameters for their particular application. Generally, if the pause time exceeds 10 seconds, the design should be modified, and the pump manufacturer should be consulted for ways to avoid this problem.

Pumps are started in some applications against a closed valve, which should be programmed to start opening when the pump reaches full speed. Variable-speed drive systems should start or operate pumps within the recommended operating range for the speed choices.

All discharge lines should be given an adequate analysis to ensure that surge pressures will not adversely affect system operation. This item is covered earlier in this chapter.

For underwater crossings and similar extended low points or inverted siphons, it is advisable to design for periodic cleaning or equip the line for blowoff to a collecting manhole.

3.0 WATER HAMMER AND SURGE

Rapid changes in fluid velocity within a pipe can lead to upsurges or down surges in pressure that can cause pipe rupture or water column separation and collapse of the pipe. *Surge* is the name given to the rapid rise or fall of pressure in a pipeline, which is caused by the conversion of kinetic energy (mV2) to potential energy (pressure) whenever fluid velocity changes. The term *water hammer* is often used synonymously with *surge*, although it is more commonly used in reference to the observable effects of surge, such as motion, vibration, or knocking noise within pipe.

Whenever a pump trips or a mainline valve or delivery valve closes, the flow is stopped, fluid velocity decreases, the upstream pressure rises due to surge, and the downstream pressure decreases due to surge. Whenever a pump is started and flow is started, fluid velocity increases, the upstream pressure decreases, and the downstream pressure increases due to surge.

When designing a new pumping station and piping system, or when making changes to an existing system that increase the flowrate or increase the speed of valve operations, it is recommended to conduct a hydraulic transient (surge) analysis to determine the magnitude of peak surge pressures and forces on piping and system components. A surge analysis can predict whether such pressures and forces may exceed allowable limits due to either high surge pressure or excessive vacuum pressure.

This chapter provides general concepts of water hammer and includes a discussion of both rigid and elastic water column theories. It also includes discussion of the causes of water hammer as well as common methods, operational techniques, and hardware for reducing the magnitude of peak surge pressures.

3.1 Surge Pressure Limits

The magnitude of acceptable surge pressure rise is governed by the applicable piping design codes. Various codes and regulatory criteria are enforced by professional engineering standards and by government agencies tasked with ensuring public safety. These codes and standards cover a range of piping materials and piping applications. United States piping codes and standards use various terms to address the pipe design pressure, maximum working pressure, or maximum allowable operating pressure (MAOP). Operational surge events occur frequently due to normal pumping station startup, normal shutdown, or changes in flowrate. Surge control methods are used to maintain routine operational surge pressure below MAOP. Infrequent surge events can occur less often, but with greater consequences. Infrequent surge events are largely due to unplanned emergencies such as power failure, pump or shaft bearing failure, sabotage, operator error, or emergency shutdown or maintenance. Surge protection methods are used to protect the integrity of pumping station piping and components during infrequent or unplanned emergencies, and to limit peak surge pressures below the maximum allowable surge pressure (MASP).

In general, the MASP for a water transmission pipeline or for piping and components within a pumping station is some specified pressure rise or offset above the MAOP. The value of MAOP governs peak pressure during steady-state operation, and planned changes in flowrate. During short-duration swings in operation, such as initial line fill, system shutdown, or maintenance procedures, the system pressure is permitted to exceed MAOP by no more than a specified surge allowance. Under no conditions is the pressure permitted to exceed MASP. Table 5.3 provides a summary of pipe codes and surge pressure allowance for many common pipe materials.

3.2 Checklist for Surge Problems

There are numerous events and conditions associated with pipeline system design of a pumping station that indicate a need for surge analysis. These events and conditions include the following:

- Electric power outages to pump motors or controls.

TABLE 5.3 Surge Pressure Allowance

Pipe Material	Applicable Code	Applicable Manual	Surge Pressure Allowance Above PC*or Max WP**
Ductile Iron Pipe	AWWA C-150	AWWA Manual M41	100 psi
Steel Pipe	AWWA C-200	AWWA Manual M11	50% of PC
Concrete Cylinder Pipe (CCP) with steel plate	AWWA C-301 AWWA C-304	AWWA Manual M9	MAX (40 psi, 40% of PC)
Concrete Cylinder Pipe (CCP) with steel rebar	AWWA C-303 AWWA C-304	AWWA Manual M9	50% of PC
Poly-Vinyl Chloride (PVC)	AWWA C-900 AWWA C-905	AWWA Manual M23	Occasional Surge, WP + $P_{OS} < 1.60 \cdot PC \cdot F_T$*** Recurring Surge, See AWWA C-905 and M23
High Density Polyethylene (HDPE)	AWWA C-906	AWWA Manual M55	100% of PC
Fiberglass Reinforced Pipe (FRP)	AWWA C-950	AWWA Manual M45	50% of PC

*PC = Pressure Class of Pipe
**WP = Maximum Working Pressure
***P_{OS} = Occasional Surge Pressure
***F_T = Temperature Coefficient

- Standby power systems cannot respond to power outages fast enough to mitigate pressure surges caused by a pump trip.
- The length of the pipeline gives a cycle period for pressure surges that is greater than a reasonable flow interruption time.
- Reverse flow through pumps.
- Forward flow through pumps after pump shutdown, such as in low head systems.
- System pressure in the pipeline is high in relation to pipe design strength before surge consideration.
- System pressures in the pipeline are low and vapor cavity formation is possible because of significant elevation change (10 m [33 ft] or more over a short distance).
- There are dead ends in the piping system conducive to pressure amplification through resonance.

- High pipeline velocities subject to rapid fluctuations.
- Pumping system is being changed to accommodate much larger flowrates within an existing piping system.
- Rapid release of trapped air generates a pressure upsurge.
- A pipeline ruptures for any reason, causing water hammer.
- Control valve and check valve closing and opening times.
- Repetitive stress from pressure surges, resulting in fatigue failure of the pipe regardless of pipe material.

In general, it is typically a good idea to conduct a hydraulic surge analysis on any new system design or with any improvements that significantly increase the flowrate or fluid velocity within an existing system, unless the cost of totally replacing the system is less than the cost of conducting a surge analysis.

3.3 Rigid Water Column Theory

Rigid water column theory (RWCT) provides some basic simplifying equations to quantify surge pressure rise without having to deal with an actual hydraulic transient analysis. The magnitude of peak surge pressures determined by rigid water column theory are not accurate for all situations. In fact, the limitations of RWCT make such calculations valid for only a subset of all real-life surge events. Rigid water column theory includes the following assumptions, which are known to be inaccurate in many cases, but greatly simplify the actual calculations involving elasticity of the fluid and the pipe and vapor cavity formation:

- Piping is rigid. Pipe has no elasticity, and it does not expand or contract due to changes in line pressure.
- Fluid is treated as being incompressible, except when determining the speed at which pressure waves travel in a pipe. Line pack is assumed to not occur in a pipeline upstream of a valve closure.
- Line pressure remains above fluid vapor pressure. Vapor cavities do not form regardless of line pressure reaching the fluid vapor pressure. There is no possible mechanism to create a spike in pressure due to the collapse of a vapor cavity.
- Fluid velocity at all locations within the pipe is the same whenever flow interruptions occur.

Figure 5.3 illustrates a scenario in which there is a pipeline of constant diameter transporting water from an upstream pump to a downstream

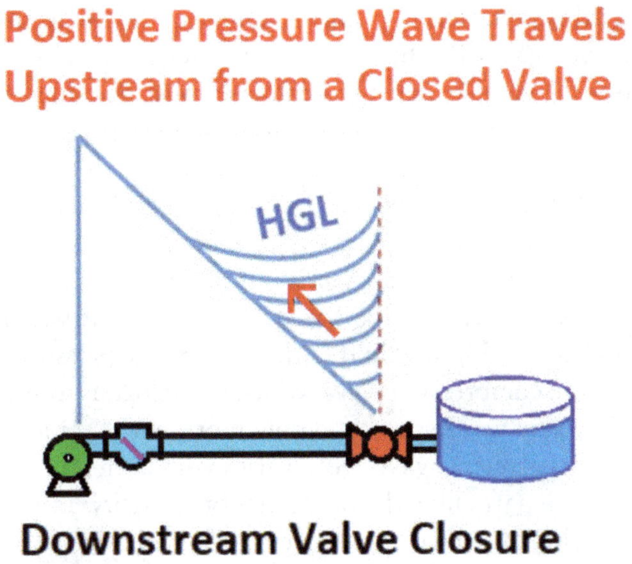

FIGURE 5.3 Hydraulic Grade Line Following a Downstream Valve Closure

delivery point, and there is a valve near the delivery point that can be opened, closed, or throttled to control the flow.

If the flow is shut off by closing the valve, then the kinetic energy of the moving water is converted into potential energy, which is manifested as a rise in pressure. The rising pressure creates an apparent surge wave, which travels upstream at the speed of sound within the fluid within the pipe. The surge wave will eventually reach the upstream pump and be reflected from the pump wet well or closed check valve and return downstream. Upon arrival at the closed downstream valve, the reflected surge wave will be reflected upstream. Each time the surge wave is transmitted through the pipeline, the magnitude of the surge wave is reduced due to friction losses within the inside pipe wall. The surge wave eventually dissipated due to line friction.

The peak rise in surge head, ΔH, that is caused by the stoppage of flow, can be calculated by the Joukowsky equation, which is derived from the equations for conservation of mass and conservation of momentum:

$$\Delta H = \frac{-a}{g} \Delta V \tag{5.12}$$

Where

a = acoustic wave speed, (m/s) or (ft/sec);
g = gravitational acceleration, (9.8 m/s^2) or (32.2 ft/sec ^{2}); and
ΔV = change in fluid velocity (V final − V initial), (m/s) or (ft/sec).

Note that this equation contains a minus sign, which implies that as fluid velocity decreases the head increases and as velocity increases the head decreases. Note also that the head rise must be added to the initial steady-state head on the upstream side of the closed valve to determine the resulting surge head.

Although RWCT tries to ignore the concept of elasticity, the bulk modulus of elasticity of the fluid and the elastic modulus of the pipe are significant parameters in determining the magnitude of acoustic wave speed (a) in Equation 5.12. Acoustic wave speed is the speed at which a pressure pulse is transmitted within the fluid in the pipe. Surge pressure waves travel through water as a compression wave, which is analogous to sound traveling through air. Acoustic wave speed is synonymous with "sonic velocity." The numerical value of acoustic wave speed varies with both fluid properties and pipe properties—in particular, the properties of elasticity. Table 5.4 provides the elastic modulus of various pipe materials and the range of wave speed when transporting water.

From Table 5.4 it is apparent that steel pipe has the highest wave speed compared to other commonly used pipe materials. This implies that if a surge analysis is to be conducted on a pumping station and pipeline system, in which the pipeline material is not yet specified, then it would be a conservative assumption to use the properties of steel pipe within the surge analysis model. Any transient simulation results based on steel pipe would be conservative if the future installed pipe is some material other than steel.

Acoustic wave speed is reduced in magnitude by the presence of air bubbles or dissolved gases within the water. This means that raw water from lakes or rivers, containing dead fish and decaying plant matter, will have reduced wave speed compared to potable water. This reduction in wave speed causes a proportional decrease in surge pressure rise within piping located upstream of a closing valve.

TABLE 5.4 Modulus of Elasticity and Wave Speed for Various Pipe Materials

Pipe Material	Elastic Modulus (1000 psi)	Range of Wavespeed (ft/sec)	Elastic Modulus (GPa)	Range of Wavespeed (m/s)
Steel	30 000	2800 to 4450	206.9	853 to 1356
Ductile/Cast Iron	24 000	3200 to 4300	165.5	975 to 1311
Concrete Cylinder Pipe	4420	2900 to 4000	30.5	884 to 1219
PVC	425	1100 to 1600	2.9	335 to 488
HDPE	155	700 to 1250	1.1	213 to 381

The magnitude of the rise in pressure, ΔP, due to surge is calculated based on the surge head rise and the density of the fluid as follows:

$$\Delta P = \Delta H * \rho * z \tag{5.13}$$

Where

ΔH = rise in hydraulic head, (m) or (ft);

ρ = density of fluid, (kg/m^3) or (lb/cu ft); and

z = units' conversion factor, to achieve (kPa) or (psi).

If one conducts a cornucopia of calculations of Equations 5.1 and 5.2 for a range of velocity change from zero to 3.048 m/s (10 ft/sec) and a range of acoustic wave speeds from 304.8 to 1 371.6 m/s (1000 to 4500 ft/sec), the results can be summarized graphically, as shown in the Figures 5.4 and 5.5. Velocity values on the X-axis represent the initial steady-state fluid velocity before valve closure, which stops the flow.

The rise in transient head (and associated rise in pressure) caused by surge must be added to the initial steady-state head (and pressure) to determine the maximum value of surge head (and surge pressure), as follows:

$$H\ (transient) = H\ (steady\ state) + \Delta H\ (surge) \tag{5.14}$$

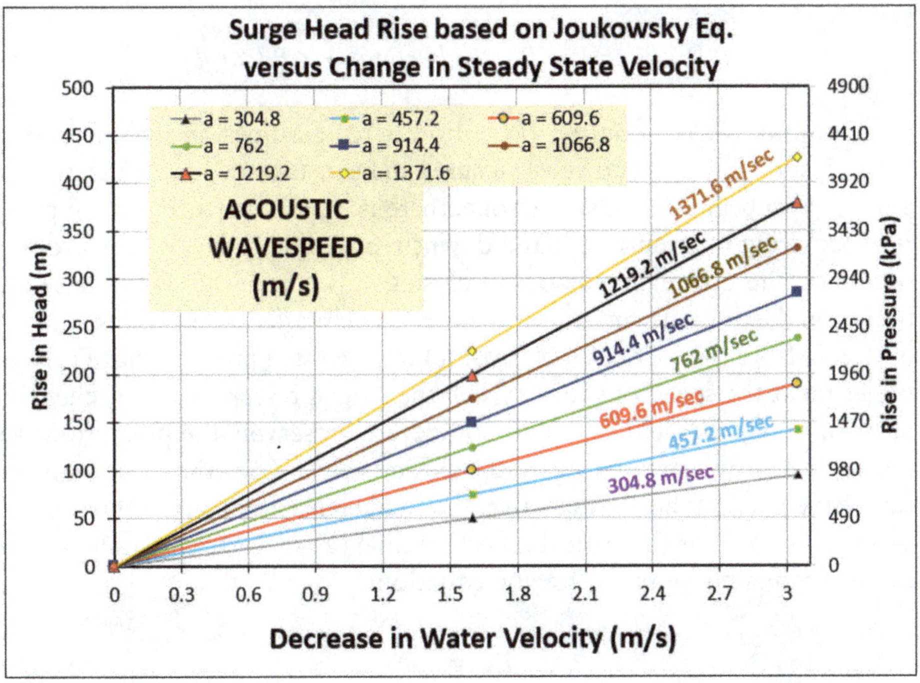

FIGURE 5.4 Surge Head Rise and Pressure Rise for Water Velocity Change (International System of Units)

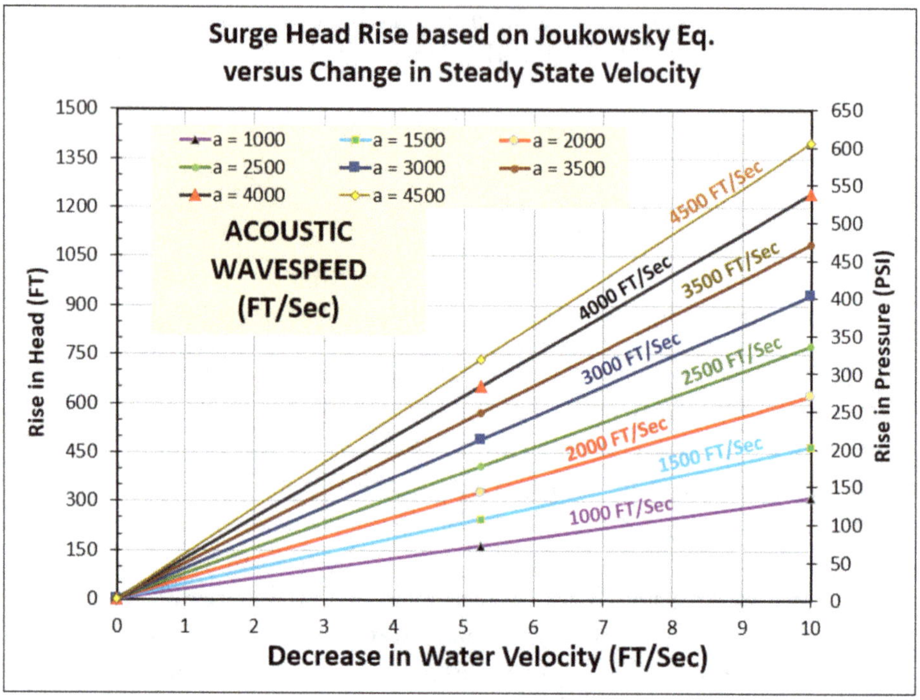

FIGURE 5.5 Surge Head Rise and Pressure Rise for Water Velocity Change (English Units)

$$P\ (transient) = P\ (steady\ state) + \Delta P\ (surge) \tag{5.15}$$

The Joukowsky equation (Equation 5.12) assumes an instantaneous valve closure. In reality, no valve closure can be instantaneous. At the instant after a valve begins its closure stroke, there is a slight perturbation in pressure occurring both upstream and downstream of the valve. A pressure rise occurs on the upstream side of the closing valve, and a pressure decrease occurs on the downstream side of the valve. The pressure rise impulse travels upstream from the valve as it closes, and the resulting pressure wave travels upstream at the speed of sound. When the surge pressure wave reaches the upstream boundary, either a pump or elevated reservoir, the pressure wave reflects back downstream toward the closing valve. When the reflected pressure wave reaches the closing valve, the total elapsed time from the start of closure is known as the critical reflection time (T_{CR}). The magnitude of T_{CR} can be calculated by the following equation:

$$T_{CR} = \frac{2L}{a} \tag{5.16}$$

Where

T_{CR} = critical reflection time, (seconds);
L = length of pipe between upstream boundary and closing valve, (m) or (ft); and
a = acoustic wave speed, (m/s) or (ft/sec).

For any valve closing time less than or equal to the T_{CR}, the upstream surge pressure achieved at the valve during the closure process will be the same as that calculated by the Joukowsky equation. If the valve closing time is greater than the T_{CR}, then the valve remains partially opened when the reflected wave returns to the valve. This means that the partially open valve will behave as an open flow path for pressure relief.

If the predicted peak transient pressure from Equation 5.16 exceeds the maximum allowable surge pressure, then surge mitigation is necessary. However, the magnitude of surge pressure predicted by Equation 5.16 is not necessarily the largest possible surge pressure.

3.4 Pressure Wave Velocity

If the assumptions and simplicity of RWCT are scrapped and replaced by the more detailed and accurate reality of elasticity within both the fluid and the pipe material, then a realistic pressure wave velocity can be determined. The velocity of the moving pressure waves will equal the velocity of sound for the combination of transported liquid and pipe materials. The formula for acoustic wave speed of pressure pulses within a fluid within a pipe is provided as follows:

$$a = \frac{\left(\frac{K}{\rho}\right)^{1/2}}{\left[1 + \left(\frac{K}{E}\right)\left(\frac{D}{t}\right) \times c\right]^{1/2}} \tag{5.17}$$

Where

a = sonic velocity, (m/s) or (ft/sec);
K = bulk modulus of elasticity of fluid, (kPa) or (psi);
ρ = density of liquid, (kgm/m^3) or (lbm/cu ft);
E = Youngs modulus of pipe material, (kPa) or (psi);
D = pipe diameter to neutral axis, (m) or (ft);
t = wall thickness of pipe, (m) or (ft); and
C = constraint factor based on method of pipe restraint, (dimensionless).

It must be noted that a unit conversion constant is required in the above equation to achieve units of (length/time) within the numerator. The denominator is dimensionless.

3.5 Negative Surge Pressure

Whenever a pumping station experiences a power failure, the pump speed spins down based on inertia (Figure 5.6). The liquid column continues to move downstream through the pipeline. The HGL decreases as the liquid column continues moving downstream. If the HGL coincides with the physical elevation of the pipeline, then the internal line pressure is zero. If the HGL falls to about 10 m (33 ft) below the physical elevation of the pipeline, then the internal line pressure decreases to water vapor pressure. At this point, the water flashes to steam vapor. A vapor cavity is formed as the same mass of water now fills a much larger volume of pipe due to the larger specific volume of steam.

3.6 Operational Changes to Reduce Water Hammer by Design

The potential ill effects of surge can be reduced or eliminated by considering the causes of surge when designing a new pumping station or pipeline

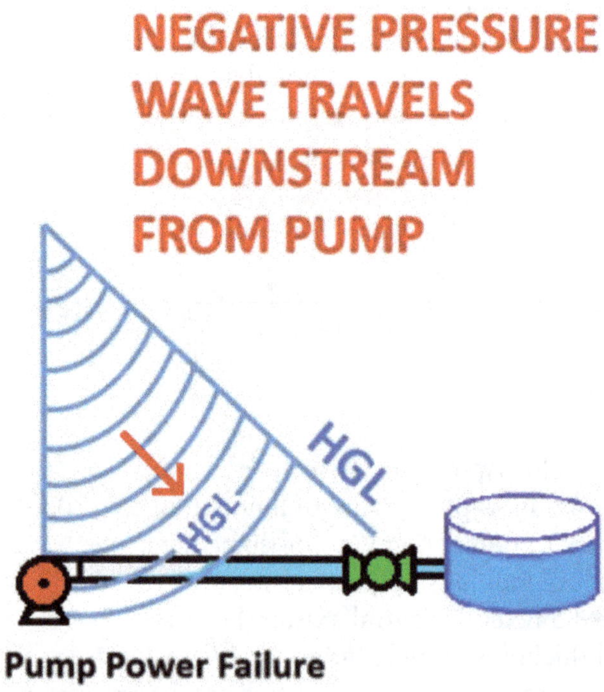

FIGURE 5.6 Hydraulic Grade Line Following an Upstream Pump Trip

system. Because the primary cause of surge is the change of fluid velocity, it may be feasible to change the design operational fluid velocity by selection of a larger pipe size to accommodate the design flowrate. It might also be feasible to specify a higher design pressure of pipe to accommodate predicted peak surge pressures. Pipe size and pipe pressure class are obvious cost issues, and a financial analysis would be needed to justify such changes in pipeline parameters.

If the pipe size and the peak flowrate are both fixed parameters, then it may be feasible to reduce the magnitude of peak surge pressure by slowing down the rate of a mainline or delivery valve closure.

If a physical system is already experiencing water hammer or surge-related issues, then a hydraulic surge model can be used to determine the maximum flowrate that can be achieved operationally, which does not cause excessive surge pressures following pump trip, valve closure, or other routine operations. Transient simulations with a surge model can be used to determine if pumping station startup and shutdown procedures are acceptable. Transient simulations of a surge model can be used to determine valve closure timing, surge relief valve setpoint pressure, and other parameters needed to prevent unacceptable surge pressures. If operational changes in the system have been simulated without success in mitigating unacceptable surge pressures, then the surge model can be configured to test the installation of various hardware components to mitigate surge.

3.7 Surge Attenuating Devices

Figure 5.4 presents and illustration of hardware equipment and methods to reduce surge pressure waves, the components of which include the following:

- Controlled closing check valve or pump discharge control valve
- Combination air vent and vacuum break valve (CAV) in the discharge line
- Surge relief valve (SRV) or pressure relief valve
- Pressurized surge tank
- Surge tower
- Pump bypass check valve
- Flywheel disc on pump shaft

3.8 Closing Check Valve or Pump Control Valve

A positive method of avoiding pressure surges on pump startup and shutdown is to control the rate of opening and closing the discharge check valve or, alternately, a pump discharge control valve (Figure 5.7). The duration

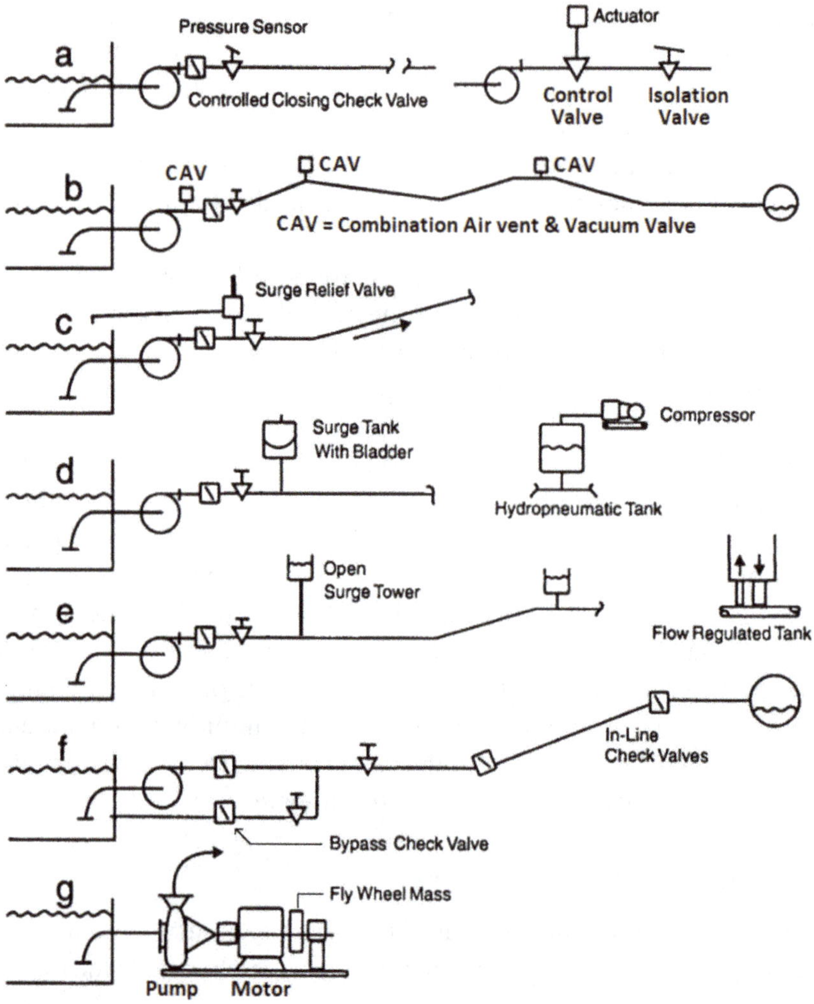

FIGURE 5.7 Pumping Station Surge Reduction Methods

for completion of valve closure must be longer than the T_{CR} for a surge wave to be transmitted downstream to a hydraulic boundary condition and reflect upstream to the pump.

In most cases, during startup of a centrifugal pump, the discharge check valve is closed or, alternately, the discharge control valve is significantly throttled. It is also possible to control the surge during startup by pumping through a smaller diameter bypass line around the check valve or control valve until stable flow is established above the minimum continuous stable flow specified on the manufacturer's pump performance curve. As the pump starts, the operating point is initially at zero (or near-zero) flow and shutoff head. As the discharge valve opens slowly, the operating point gradually slides down the head-flow curve to reach its normal operating point.

Under normal shutdown procedures, the pump discharge valve closed at a predetermined and field-adjustable rate before stopping the pump. Slowly opening and closing pump discharge check valves or control valves provides an effective way to control surge pressures during normal pump startup and shutdown operations. However, during a power failure event, the pump will spin down suddenly based on inertia, with a sudden disruption of flow. It is recommended to have valve actuators (pneumatic or hydraulic) available to slowly close the discharge valve following a power failure. Backup pneumatic or hydraulic power should have adequate reservoir capacity to accommodate the complete valve closure cycle.

3.9 Combination Air Vent and Vacuum Break Valves

It is recommended that a CAV be installed on the discharge piping of each pump within a pumping station. It is also recommended that a CAV be installed on the common discharge header pipe of a pumping station having two of more pumps configured in parallel. The primary purpose of CAVs is to allow air ingress to the piping to try to maintain positive pressure whenever the pipeline hydraulics contain vacuum pressure during system operation. Another purpose of CAVs is to vent air from the piping during line fill operations and during normal operations whenever air ingress occurred to displace vacuum conditions.

It is recommended that CAVs be installed along the pipeline at relatively high points in elevation and at changes in pipe slope that are concave downward. American Water Works Association's (2016) design manual M51 identifies a more complete list of locations along the pipeline, at which CAVs are recommended. The design manual includes recommended CAV spacing criteria for long stretches along the pipeline, which are horizontal and have a shallow constant slope. Computer simulation results of expected hydraulic transient events in pipelines may indicate that some CAVs at recommended installation locations will never open during or following system startup or shutdown, or normal swings in operation. However, such CAVs may be needed to function in the event of line breaks caused by external damage from non-hydraulic sources.

Air ingress and air venting functions of CAVs are available as either a single valve body or as a smaller air vent valve attachment to a larger vacuum valve body. Combination air vent and vacuum break valves may use metallic bulb floats on connecting rods or linkages or multiple floating discs to change their operation from air ingress to air venting. Some CAV designs use check valves or special internal flow paths to minimize surge effects when the CAV opens or closes.

There are two predominant methods of determining appropriate size of CAVs for air ingress purposes. The simplest method uses an explicit formula to determine liquid flowrate during system drain down and either trial and error or a lookup table or figure to determine CAV valve body size. The explicit calculation involves the Chezy formula, which has input parameters of liquid flowrate during drain down, slope of the pipeline, pipe diameter, and a Chezy coefficient. This method has no relevance to the vacuum pressures expected during transient events such as pump power failure or upstream valve closure. The other method of sizing CAVs involves computer simulation of the transient event and determination of the worst-case vacuum pressure and calculation of the maximum air ingress velocity observed during the simulation. The predicted vacuum pressure and peak air ingress velocity is then compared to the maximum allowable air ingress velocity recommended by the CAV manufacturer. If either the predicted vacuum pressure or air ingress velocity are excessive, then the CAV size is increased in the model and the transient simulation is rerun until acceptable results are achieved.

3.10 Pressure Surge Relief Valves

To relieve excess surge pressures in a pipeline, an SRV can be installed on the pump discharge pipe. The SRV is normally closed but will open when the inlet pressure exceeds a specified setpoint pressure, SPOPEN. The relieved flow is normally discharged back to the pump wet well, supply tank, pump suction header, or other suitable delivery point. Surge relief valves may be installed as a supplemental device for surge protection in addition to other surge control devices used during normal operation.

When an installed SRV initially opens, relief flow cannot begin until after the upstream pressure travels as a compression wave down the relief line, reflects off the relief discharge point, and travels back to the SRV. This time lag is an example of the T_{CR} described by Equation 5.16. The magnitude of relief flow through the SRV is a function of the flow coefficient of the SRV, the resistance to flow within the relief piping, and the pressure at the point of discharge. Surge relief valves may be designed to use either an inline flow path or an angle-style flow path in which relief flow makes a 90-deg bend.

Once an SRV opens, it will remain open until the inlet pressure falls below a specified setpoint for closure, SPCLOSE. The magnitude of SPCLOSE is typically in the ballpark of the greater of (90% of SPOPEN) or (SPOPEN – 10 psi). There should be enough offset between the magnitude of SPOPEN and SPCLOSE to prevent chatter of frequent repetitive opening and closing.

Valve manufacturers have developed methods for sizing SRVs for specific applications. These methods typically produce a conservative design. An hydraulic transient analysis can also be used to determine the appropriate valve size and pressure setpoints for opening and closing of SRVs. Computer simulation models can determine how much peak surge pressure continues to rise while the SRV is opening, and the critical reflection time is passing.

3.11 Surge Accumulator Tanks

A surge accumulator tank is a code-stamped pressure vessel that is connected to a pipeline and is equipped with a compressed gas supply to maintain a predetermined range of operating pressure within the tank. The pressure vessel may be a horizontal cylinder, vertical cylinder, or a sphere. If the pressure vessel is a cylindrical shape, then it will typically have dished endcaps, which are slightly rounded to hemispherical in shape. Surge accumulator tanks range in volume from the size of a baseball up to 189.3 m^3 (50 000 gal).

The enclosed gas within the surge tank is typically compressed air supplied from a compressor or from a compressed air cylinder. In systems that contain flammable or toxic liquids, the gas is nitrogen-supplied from a compressed N_2 cylinder. In some systems using compressed air, the air will be floating on top of the liquid and may be in direct contact with the liquid top surface. More recent designs use an internal bladder to maintain separation between the air and the liquid. The compressed air will typically fill the bladder to an initial "pre-charged" operating pressure. During swings in system operation, the volume of the compressed air will expand and contract with changes in line pressure. Ideally, the maximum change in air volume should range from 20% to 80% of total tank volume.

When a surge tank is connected to a water pumping station discharge header, it will provide makeup water into the pipeline following a pumping station power failure. The same surge tank will act as a shock absorber when a reflected surge wave comes back to the pumping station by admitting more water into the tank, which compresses the air to a higher pressure.

Surge tanks may be used in conjunction with other devices such as slow-closing and slow-opening check valves or SRVs. In some cases, the incoming and outgoing liquid lines to/from the surge tank require different flowrates and, therefore, different resistance to flow. In such cases, the incoming and outgoing lines may be different sizes with a check valve in each line before the lines merge into a single flow connection at the tank. In some cases, the incoming and outgoing flow lines may be the same line, which contains an asymmetrical orifice or American Society of Mechanical Engineers nozzle to provide the needed variation between incoming and outgoing flowrates.

The size and location of pressurized surge tanks is best determined by the results of hydraulic transient simulations conducted by a surge expert. Transient simulations may require multiple trial-and-error attempts to optimize the size and location of a surge tank. Because of the relatively high cost of code-stamped pressure vessels, it is usually worth the cost of properly sizing the optimum surge tank rather than procuring an oversized surge tank with excessive safety margins.

3.12 Surge Towers

Surge towers are open vertical tanks with the water surface exposed to atmospheric pressure and the tank bottom connected to the pipeline. If the water surface is physically open to atmosphere, as well as to birds, insects, dust storms, and other debris, then the installation of such surge towers should be limited to untreated raw water or to agricultural water systems.

A surge tower acts as an equalizer for flow variations by discharging liquid during a down surge or by filling with more liquid during an upsurge. The HGL of the pipeline, from upstream pump discharge to downstream delivery, must be capable of passing through the vertical surge tower height. The HGL of the discharge line must be low enough for an open tank to be used, and, if needed, a surge tower at a peak in the discharge pipeline elevation profile can be constructed.

If the surge tank is located close to the pumping station, it can protect the pump upon startup from discharge line inertia that may cause the pump to otherwise operate at zero flow and shutoff head. The height of the tower must account for the inertia head pressure, the static and friction head losses, and the upsurge head at the maximum combination.

There are several accessories to enhance functionality of surge towers. By installing a check valve to allow flow only out of the surge tower into the pipeline, a down surge wave can be attenuated. An altitude valve can control the filling rate to prevent overflow, or, if HGL permits, a smaller open return line can be installed.

A hydraulic transient simulation model can be constructed and executed to determine the swing in water surface elevation within a surge tower. Simulation results can confirm whether a surge tower design is appropriate and has adequate freeboard. Local ordinances will govern the use of lighting on the surge tower to alert low flying aircraft.

3.13 Bypass Check Valves

A simple arrangement for protecting a pipeline against down surge pressures are to install a bypass check valve in parallel with the pump. The check (non-return) valve would discharge in the same direction as the pump. Under normal operating conditions, the pump discharge pressure would be higher than the pump suction pressure, and the pressure difference would maintain the check valve in a closed position.

Following a normal pump stop, or even an unplanned power failure, the pressure in the pump discharge pipe would drop as the liquid column continues to move downstream. If the discharge pressure drops to below the pump suction pressure, water would be drawn through the bypass valve until the discharge and suction pressures are equalized. This process reduces the magnitude of the upsurge wave pressure.

This method of surge pressure protection can only be effective if the discharge pressure were to drop below the pump suction pressure. If the system discharge pressure is low and the discharge line velocity is high, as in a long level discharge pipeline with high flow, then bypass check valves can provide an inexpensive method of surge reduction. The suction reservoir level and volume must be adequate to fill the pipe to avoid water column separation in the discharge line. Pressure surge relief valves may still be required as supplementary surge protection.

Water could still be drawn through the pump case during the period when the delivery head is below the suction head. A controlled closing pump check valve or isolation valve would control this potential problem but could also remove the need for the bypass check.

3.14 Pump Flywheels

Following a pump trip, the rate at which the pump impeller speed decreases is governed exclusively by the combined rotational inertia of the pump impeller, motor, connecting shaft, and liquid within the pump case. In general, the pump impeller spins down rapidly, which creates a significant down surge "rarefaction" wave, which travels downstream within the pipeline. The magnitude of the down surge in pressure can be reduced by increasing the combined inertia of the pump, motor, and shaft. The inertia can be increased by installation of a large steel disc (flywheel) onto the rotating pump shaft. If a flywheel is installed, the motor must be designed to accommodate the added load during pump startup.

3.15 Surge Analysis Software

There are several different software programs/packages available in the marketplace that are suitable for conducting hydraulic surge analysis of a pumping station and pipeline system. Each of these programs must solve the partial differential equations that govern the flow of liquids in the pipe. The first differential equation, called the *continuity equation*, that must be solved pertains to the balance of mass flow in/out of a control volume with respect to space and time. The second differential equation, called the *momentum equation*, to be solved pertains to balancing the momentum of the liquid and the forces imposed on the liquid within a control volume with respect to space and time. Both of these equations, the continuity equation and momentum equation, can be expressed as partial differential equations in which the dependent variables of head (or pressure) and fluid velocity (or flowrate) are functions of the independent variables of space and time. Both equations can be solved simultaneously using various numerical methods.

Most of the marketed surge analysis software programs use the method of characteristics (MOC) for solving the differential equations that govern fluid flow in pipes. Numerical models of pumping stations and pipelines will typically use some base length, ΔX, or computation section as the smallest spatial length of pipe over which the differential equations will be solved during each timestep of a transient simulation. Each and every pipe within the numerical model must have a specified length, which is equal (or approximately equal) to some integer multiplier times the selected base length, ΔX. Each pipe will also have an input parameter to quantify the acoustic wave speed (a) at which pressure pulses travel within the fluid that is transported in the pipe. The entire model will also have a specified (and fixed value) timestep parameter (Δt) that is used for each timestep as the simulation progresses through time.

Numerical models using MOC must adhere to certain criteria to maintain stability and accuracy. The most predominant of these criteria is referred to as the Courant-Friedrichs-Lewy (CFL) stability criterion, which states that the product of wave speed (a) and timestep, Δt, must be less than or equal to the spatial base length, ΔX. Dividing both sides of the equation by ΔX results in the following equation:

$$CFL = \frac{a\Delta t}{\Delta X} \leq 1 \qquad (5.18)$$

Where

a = acoustic wave speed (length/time),
Δt = the fixed time step (time), and
ΔX = the base length of spatial distance of pipe (length).

The Courant stability criterion requires that the surge modeler pay close attention to the selection of time step, the lengths of each pipe segment, and the acoustic wave speed associated with each pipe within any surge model using MOC. The process of schematization is used to take length from one pipe and attach it to a contiguous pipe to resolve conflicts in satisfying the Courant stability criteria. Failure to adhere to the Courant stability criterion for all pipes in a surge model can lead to erroneous results.

Another numerical method that is used by several popular surge software programs is called the *wave method*. The wave method requires fewer calculations than MOC because it does not calculate pressure and fluid velocity at intermediate computation sections within long pipes. The wave method does not perform frictional pressure drop calculations at intermediate points within each pipe. The wave method will determine a linearized frictional pressure drop per unit length, and then multiply that value by the length of the pipe to determine the frictional pressure drop across the entire length of each pipe element. This process saves a significant amount of computation time on each timestep compared to MOC.

Both MOC and the wave method use explicit numerical solutions of the differential equations pertaining to mass balance and momentum with respect to space and time. Other numerical methods have been used by surge software programs that use implicit numerical solutions of the differential equations governing fluid flow. These other methods include the Galerkin method, and a "Theta weighted" solution to an implicit algorithm that can implement a "backward in time, central in space" finite difference solution, or the implicit Crank–Nicholson scheme for solution of the partial differential equations. The implicit numerical methods do not require adherence to the Courant stability criterion and, therefore, model schematization is not required. Implicit numerical models do not require a fixed timestep. The implicit numerical schemes use a user-specified pressure tolerance and a variable, but optimized value of Theta to maintain convergence and stability within each timestep.

A brief list of some of the surge analysis software programs includes the following: AFT Impulse, Bentley Hammer, InfoSurge (by Innovyze), Pipe2022:Surge (by KY Pipe), LIQT (by DNV), Synergi Pipeline Simulator (by DNV), TransAM (by HydraTek), and Pipeline Studio (by Energy Solutions Inc.). It is important that whenever a surge analysis is conducted, the surge modeler is aware of the limitations of the software and of any default values that are assumed by the software if left unchanged by the surge modeler.

Surge analysis software will have a wide range of menus and features within the graphical user interface, which can span a wide range of flexibility, user-friendly instructions, and various levels of helpful content in the user manual.

Surge analysis software using explicit methods (MOC or wave method) for solving the differential equations will offer a choice of turbulent flow frictional pressure drop equations to allow the user to choose between Hazen–Williams or Darcy–Weisbach equations for frictional pressure drop. Alternately, surge analysis software using implicit numerical methods offers only the Darcy–Weisbach equation for turbulent flow frictional pressure drop, but also uses appropriate friction factors for characterizing laminar flow and the critical flow regime between laminar and turbulent flow.

4.0 PIPING MATERIAL AND FITTINGS

The most common pipe materials used inside wastewater and stormwater pumping stations are ductile iron, cast iron, and steel. Discharge station piping and fittings are either of steel, cast iron, or ductile iron construction. The ability to fabricate complex steel shapes for station discharge piping has increased the use of steel. Steel pipe must be protected against corrosion both externally and internally. Concrete pressure pipe, ductile iron pipe, cast iron, and HDPE pipe are common on discharge lines outside the pumping station. Composite piping materials are also used. The selection of a pipe material is typically based on pipe strength, corrosion resistance, local regulations, and cost. Suction side piping and fittings in dry well–wet well station piping typically are of ductile iron.

Cast iron pipe fittings are available but are more brittle than those of ductile iron. Piping and fittings of iron can be internally coated with polyethylene, asphaltic, or cement mortar linings for corrosion protection and to maintain a low friction loss coefficient.

4.1 Pipe Joints and Connections

Cast iron piping sections typically are joined by gasketed flanged joints 100 mm (4 in.) in diameter and larger. Steel piping sections are typically joined by welded joints or by flat face flanges for connections 100 mm (4 in.) in diameter or larger. Threaded couplings and fittings are used in sizes 100 mm (4 in.) and smaller in diameter. Pipe flanges may be integrally cast or threaded to the pipe by the manufacturer.

There are several proprietary clamp-on joint fittings, which permit a level of deflection in the pipe and simplify installation and future repair or replacement. Grooved pipe and pipe fittings are used to provide leak-tight connections while allowing for some angular pipe deflection.

Sleeve or sleeve-type couplings are used to join pipes and allow axial expansion and misalignment. It is important that the joints be harnessed with restraining bolts and provided with restraints at bends.

Pipe for underground force mains can be supplied with bell and spigot joints when outside the pumping station in combination with thrust blocks. Mechanical restrained joint fittings, which use bolted retainer glands, are also used in underground piping in lieu of bell and spigot thrust block combination. Also, fusible PVC and HDPE are common joint connection methods used that incorporate high heat to melt the ends of the pipe joints and then fuse them together. Since fusible methods are similar to welded connections, the joints do not require restraining. To avoid or reduce the differential settling effect between the pumping station and the force main, a flexible pipe joint may be installed outside the station between station piping and the force main.

4.2 Pipe Supports for Static Loads

Piping systems must be structurally sound. The static support system must consider the structural strength and weight of the pipe and fittings and the jointing arrangement. Thermal expansion of the pipe must be recognized.

Static supports of concrete or metal saddles and hangers are used to support the weight of station piping. Base-mounted elbows frequently are used along with lateral supports on vertical pipes. The spacing of supports and their application have been calculated and presented by the manufacturers (Grinnell Corporation, 1987). The design of concrete supports should provide for axial pipe movement without progressive wear of the pipe exterior wall at the support.

Vibration from surge pressures or other hydraulic transients imposes unusual loads on pipe supports, making it prudent to design conservative support systems and provide for both weight and axial-lateral loads. Support systems must be designed for possible over-pressurizing in required pipe testing. Computer programs, such as Surge—A Computer Analysis of Transient Flow, are available for some or all of this work.

Expansion joints, sometimes used at pump flanges, typically are constructed of two metal flanges with a single arch nonmetallic tube connecting the flanges. The use of expansion joints with proper anchoring devices will reduce, but not eliminate, vibration to or from the pumps and piping system. Specifications should call for proper expansion joint installation (Fluid Sealing Association, 1980), though their use should not be an excuse to use the fitting to accommodate misaligned piping.

An expansion joint should be installed with an anchor point between the expansion joint and the pump. The designer should realize that exposed joints must be harnessed or blocked to resist large thrust forces developed with the expansion joint. Figure 5.8 illustrates one method of avoiding the transference of these forces to the pump (Fluid Sealing Association, 1980).

4.3 Pipe Supports for Dynamic Loads

When a pipe filled with liquid encounters a change in direction, as at a bend, unbalanced thrust forces (pressure and momentum forces) must be balanced with anchored supports called *thrust blocks*. Thrust blocks for underground pipes typically are concrete structures designed for the thrust loads and soil conditions. In-station piping must also be analyzed. The basic relationships for treating these loads are shown in the formulas below.

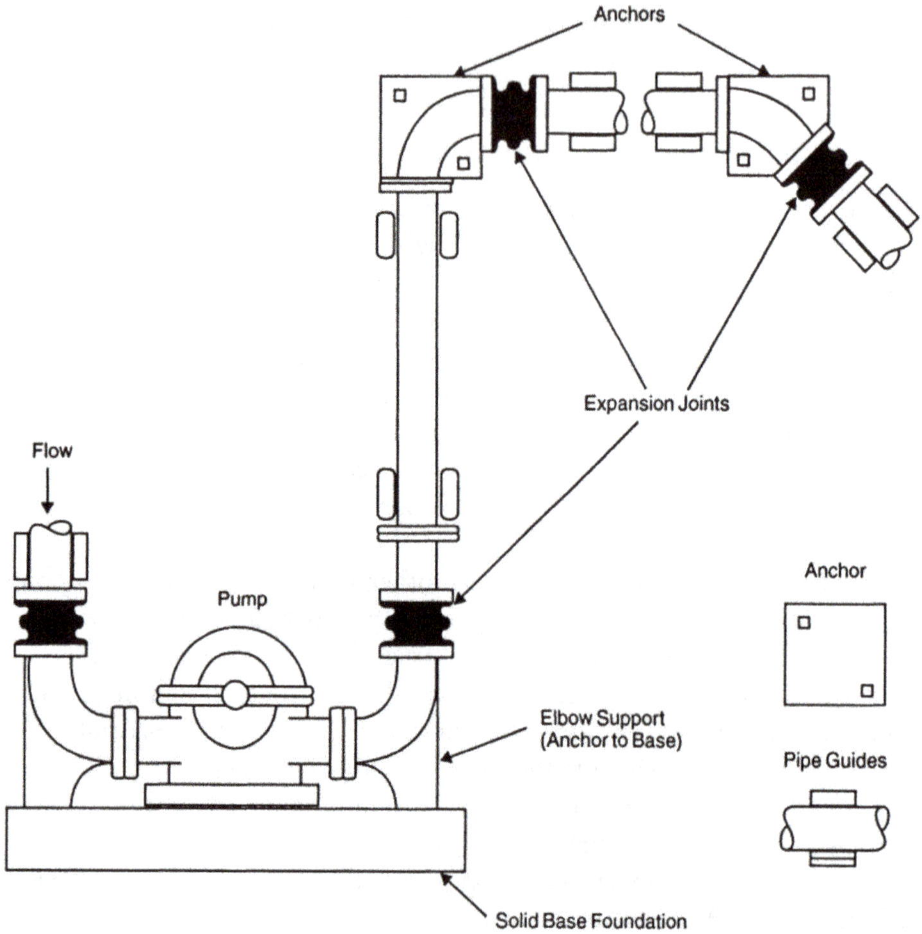

FIGURE 5.8 Typical Piping Layout Using Expansion Joints When Equipment and Piping Are Properly Anchored

A summation of the forces relating to internal pressures (R) and the forces relating to changes in linear momentum (M) are

$$R_x = P_1A_1 - P_2A_2 \cos\theta \qquad (5.19)$$
$$R_y = P_2A_2 \sin\theta$$
$$M_x = GQ\,[V_2 \cos\theta - V_1]$$
$$M_y = GQ\,[V_2 \sin\theta]$$
$$R^2 = R_x^{\,2} + R_y^{\,2}$$

Where

P = internal pressure,
A = area,

θ = bend angle degrees,
G = mass density (constant),
Q = flow rate (constant), and
V = velocity.

In the most general case, areas A_1 and A_2 are different. Figure 5.9 shows the formulas to calculate the reaction load for the most common cases but does not consider momentum changes. The largest area of a bend should be used to yield a conservative thrust value.

The change in linear momentum contribution to the thrust load reaction is minimal in most cases. The contribution from linear momentum changes increases with velocity change of the fitting.

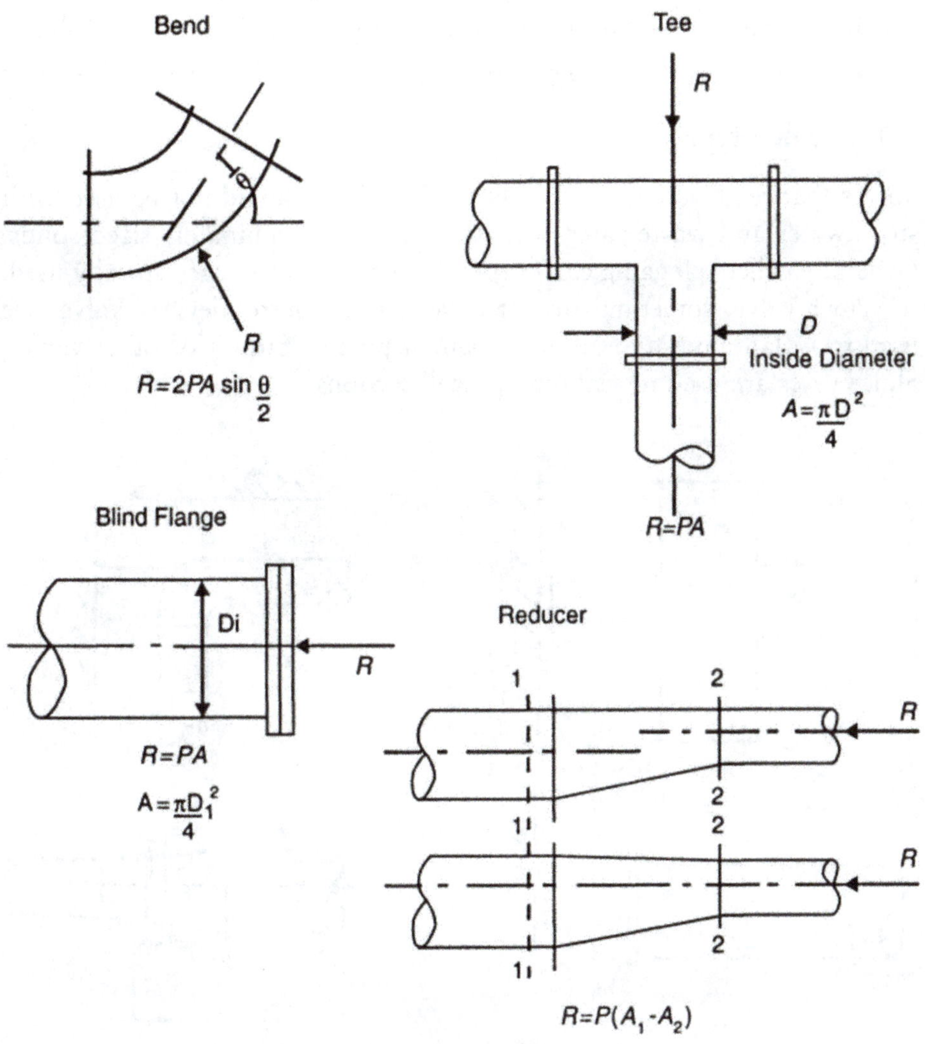

FIGURE 5.9 Pipe Thrust Force Formulas

For in-station piping, thrust restraints are made of steel. Concrete is typical for underground piping. Figure 5.10 shows some general in-station thrust blocking and restraining arrangements. As an alternative to blocking, restrained joint piping can be used to resist thrust forces.

For design of thrust block for underground piping, soil mechanics and pipe material determine much of the design. Figure 5.11 illustrates underground, vertical, and horizontal thrust blocking and restraint designs.

5.0 VALVES

Pumping stations use valves to isolate a pump, prevent backflow through the pump, isolate the valve that prevents backflow, divert or bypass flow, isolate segments of lines for servicing, and surge control. Figure 5.12 shows a general schematic for a typical pumping station valving arrangement.

5.1 Valve Types

Valves that readily clog, such as butterfly valves, should not be used with stormwater and wastewater, which contain indiscriminately sized solids. Generally, where clogging can occur, plug or ball valves are typically used.

Pinch valves sometimes are effective as flow control devices. Valves are used to isolate lines and permit repair of pipe or fittings or other valves. Sluice gates are used to seal off wet well sections.

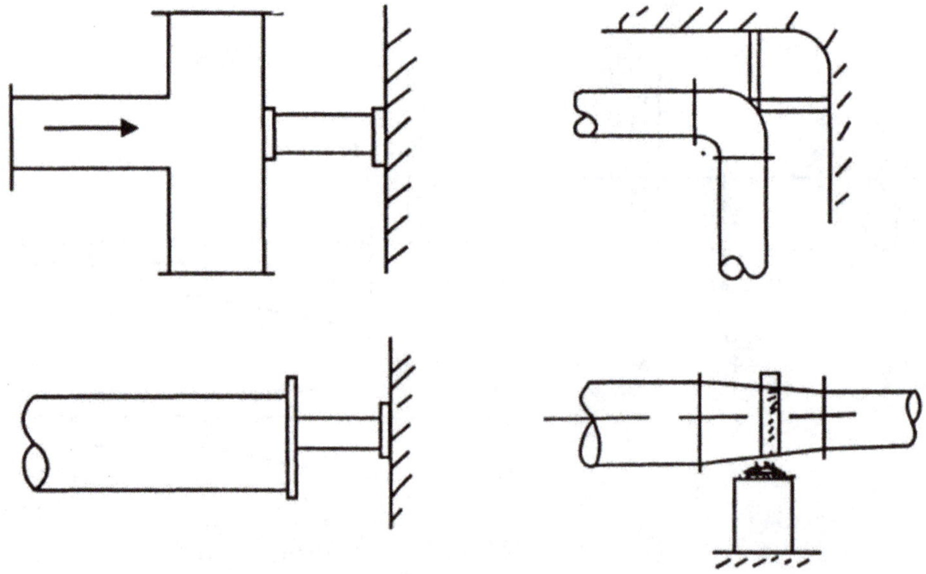

FIGURE 5.10 Station Piping Thrust Restraints

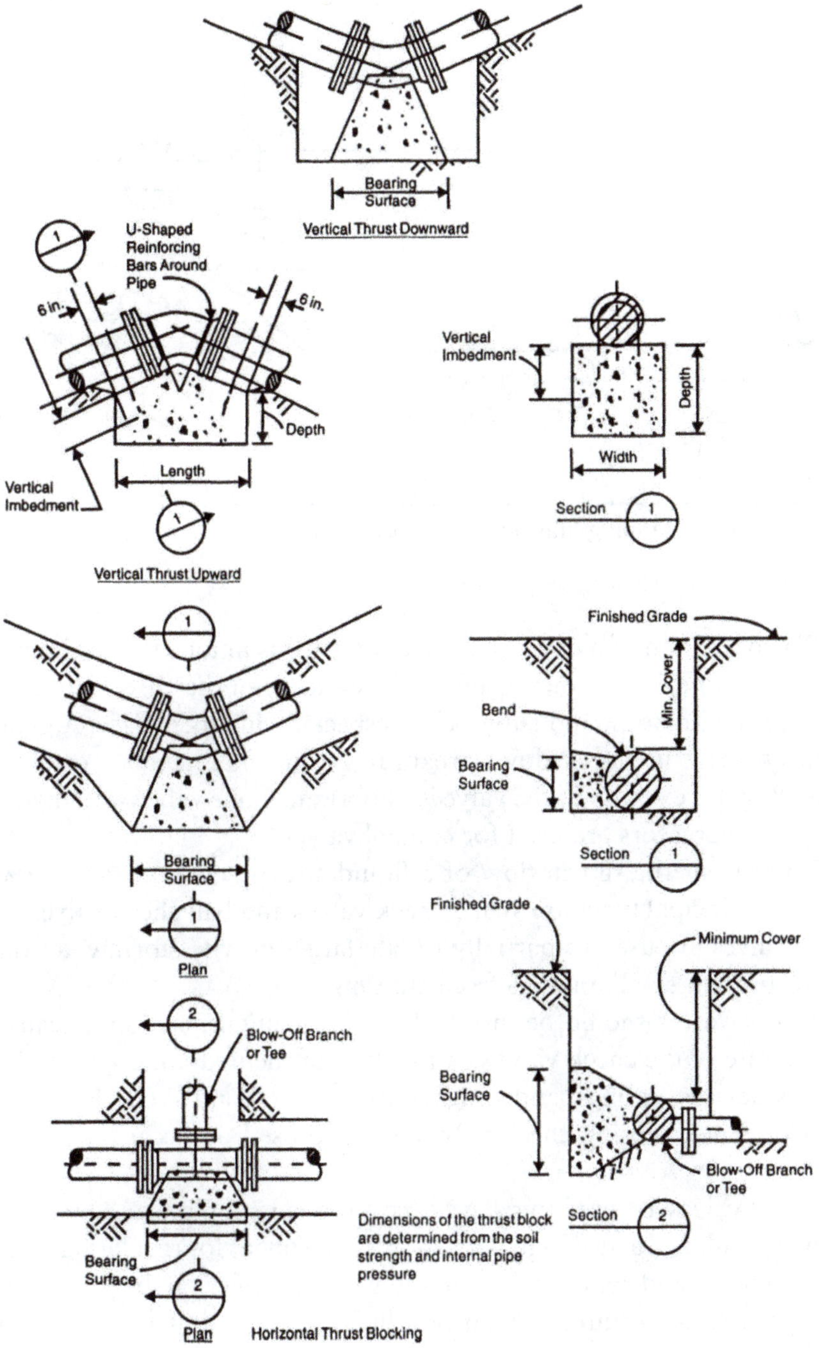

FIGURE 5.11 Underground Piping Thrust Restraints (in. × 25.4 = mm)

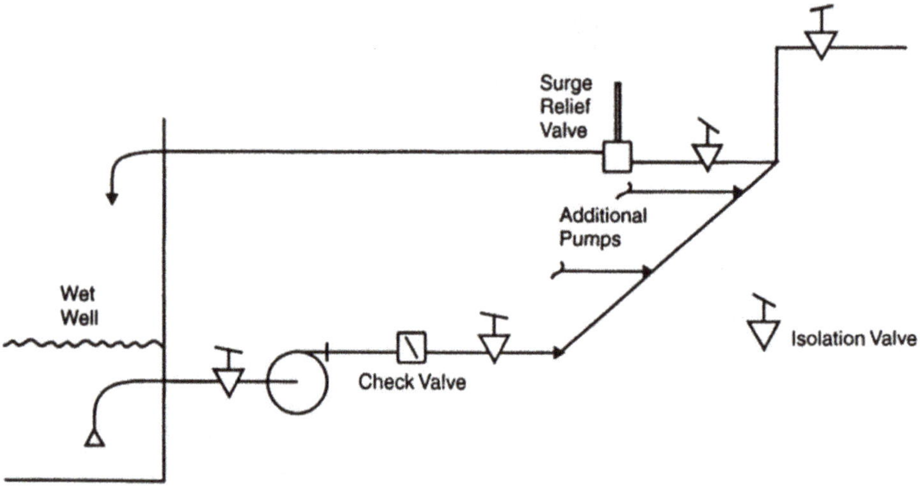

FIGURE 5.12 Pumping Station Valve Locations

The operation of a cone valve (Figure 5.13) is unique in that it involves lifting the cone, which then rotates 90 deg to open the valve. Cone valves are used for isolation and surge relief, especially in larger sizes. A guard or isolation valve is placed downstream to permit maintenance of the cone valve. Regular exercise of the valve is important. Cone valves with hydraulic or electric actuators are used for control valves.

To prevent the return flow of a liquid stream, check valves are used. The two principal types are swing check valves and ball check valves. Pinch check valves are used to partially isolate larger gravity stormwater outlets and to prevent small animals from entering.

Check valves should be installed in a horizontal position. Because the closing rate of the check valve is a function of the hydraulic system design, various devices, springs, and weights are used to control check valve closing. Controlled closing check valves are also used on occasion to prevent slamming.

Relief valves are used to relieve excess pressure in pipes and are typically of an angle globe valve design with spring adjustment for regulating pressure.

Air release and combination air and vacuum valves can be placed at high points to release accumulated air or admit air under down surge operating conditions.

Trapped air at pipeline high points can increase power requirements and the potential for pressure surges. Manual air release valves are a possible solution to this problem.

Trapped gases can be high in hydrogen sulfide and cause considerable problems with odors. Also, the orifice size in the valve may cause high

FIGURE 5.13 Cone Valve (Partially Open)

velocities, which can produce elevated sound levels in the vicinity. Smaller accessory valves may be of globe or gate design.

5.2 Valve Actuation

Smaller valves can be opened and closed manually by adjustment of an operator. Line pressure and valve size may require the use of mechanical actuation devices (gears and chain drives) to facilitate opening and closing.

Powered actuators are used for larger valves to provide remote operation or to positively open or close valves. Actuators are powered by electric, pneumatic (compressed air), or hydraulic pressure. Actuators can simply open or close a valve or be used to position a valve. The rate of opening and closing can be made adjustable.

Pumping systems frequently use power-actuated valves as an isolating device. Where ordinary swing-type check valves are not suitable, controlled closing and opening devices may be installed on the swing check. A plug or ball valve can also be used.

5.3 Valve Applications

In wastewater and stormwater pumping stations, valves are used to interrupt or control flow, vent gases or relieve a vacuum condition, or relieve excess pressure from a system.

An important application is pump check service. During a power failure, the actuator (pneumatic or hydraulic) can operate the check valve in a predetermined emergency cycle if supplied with emergency air or liquid under pressure.

Valves should not be mounted directly on the suction side of a pump because the valve port can create turbulence in the pump suction, leading to serious problems.

Surge relief valves should be sized with minimum pressure drop in the discharge line. Some installations use two surge relief valves, frequently of different sizes, operating in parallel with slightly different pressure-relief settings. The port area of the two valves must equal a port area of the larger single valve required.

6.0 FLOW METERS

In addition to establishing flowrate for billing and remote flow control, flow is measured at pumping stations to test pumping systems precisely. Even under the best planned conditions, field tests are difficult to perform accurately. A properly installed and recently calibrated flow meter can add accuracy to the field test.

7.0 REFERENCES

American Water Works Association. (2016). *Air valves: Air-release, air/vacuum and combination* (2nd ed.; AWWA Manual of Water Supply Practices M51).

CEComp. (1986). Micro Hardy Cross Computer Program.

Crane. (1974). *Flow of fluids through valves, fittings and pipe* (Technical Paper No. 410).

Fluid Sealing Association. (1980). *Technical handbook* (5th ed.). Fluid Sealing Association, Rubber Expansion Joint Division.

Great Lakes—Upper Mississippi River Board of State Public Health and Environmental Managers. (1990). *Recommended standards for waste water facilities*. Health Education Services.

Grinnell Corporation. (1987). Hanger standards.

Heald, C. C. (1992). *Cameron hydraulic data* (17th ed.). Ingersoll Rand.

Hydraulic Institute. (1979). *Hydraulic Institute engineering data book*. Hydraulic Institute.

Hydraulic Institute. (1983). *Hydraulic Institute standards for centrifugal, rotary & reciprocating pumps* (14th ed.).

Krebs, J. R. (1990). *KCS pump—Centrifugal pump selection by computer*. Krebs Consulting Services.

Submersible Wastewater Pump Association. (1986). *Submersible sewage pumping systems handbook*. Lewis Publishers.

University of Kentucky. (n.d.). KY Pipes Computer Program. University of Kentucky, Department of Civil Engineering.

Water Environment Federation. (2020). *Odor Emissions and Control for Collections Systems and Water Resource Recovery Facilities* (2nd ed.; WEF Manual of Practice No. 25).

Wood, D. J. (1991). *Waternet-flow distribution network analysis*. University of Kentucky, Department of Civil Engineering.

8.0 SUGGESTED READINGS

Chaudrey, M. H. (1987). *Applied hydraulic transients*. Van Nostrand Rheinhold.

Flygt. (1993). Design recommendations for pumping stations with large submersible Flygt centrifugal wastewater pumps.

Karassik, I. J., Krutzsch, W. C., Fraser, W. H., & Messina, J. P. (1976). *Pump handbook*. McGraw-Hill.

Krebs, J. R. (1993). *Pump* selection and system design with the microcomputer. Lewis Publishers.

Metcalf & Eddy. (2014). *Wastewater engineering: Collection and pumping of wastewater*. G. Tchobanoglous (Ed.). McGraw-Hill.

Parmakian, J. (1963). *Water hammer analysis.* Dover Press.

Prosser, M. J. (1977). *The Hydraulic design of pump sumps and intakes.* British Hydromechanics Research Association.

PUMP-FLO. (n.d.). PUMP-FLO: A Pump Selection Program, Engineered Software.

Streeter, V. L., & Wylie, E. B. (1967). *Fluid transients.* FEB Press.

Thorley, A. R. D. (1992). *Fluid transients in pipeline systems.* D&L George, Ltd., Barnet Herts.

University of Kentucky. (n.d.). *Surge—A computer analysis of transient flow.* University of Kentucky, Department of Civil Engineering.

U.S. Environmental Protection Agency. (1979). *Sludge treatment and disposal* (EPA-625/1-79-011).

6

Electrical Design

Electrical design of wastewater and stormwater pumping stations must account for all electrical loads and provide cables and service equipment that meet local standards and codes. Electrical system design must also provide cables for interfacing the components of the instrumentation and control system.

An electrical system can range from a simple fractional kilowatt motor plugged into a convenience outlet to a complex group of pumps with megawatts of connected load. Local jurisdictions have final authority for the enforcement of standards and codes used in electrical system design.

1.0 RELIABILITY AND REDUNDANCY

Reliability and redundancy requirements of electrical system design are covered in the codes and standards. The National Electrical Code (NEC) covers requirements for the electrical system conductor sizing, grounding, and motor circuit protection; installation requirements for hazardous areas; and installation requirements for standby generation. The *National Electrical Safety Code* (National Fire Protection Association [NFPA], 2020a) covers requirements for substation and overhead line designs. The NFPA's *Standard for Fire Protection in Wastewater Treatment and Collection Facilities* defines hazardous areas for pumping stations (NFPA, 2020c). The Institute of Electrical and Electronic Engineers (IEEE) 3006 Standards: Power System Reliability (n.d.-b) discuss the reliability of industrial and commercial power systems.

The local jurisdiction has final authority on standards used in electrical design for any project and must be consulted about their requirements for the electrical system. Also, the means of service to the facility by the local utility company must be incorporated into the design.

The utility's reliability is reflected in its outage history for the service line to the facility: what happens if power to the facility fails, and what facilities does the owner or utility have for handling such an occurrence? An outage will not be a problem if storage capacity is built into the system or if there is available standby pumping via portable pumps or standby generators.

For other facilities, provisions are needed to ensure pumping is not lost. The level of reliability and redundancy needed will be dictated by the political realities of the facility's location. The installation of a standby pump sometimes will be sufficient for some facilities. For others, standby generation or a second utility source will be needed. In critical locations, the facility may require redundant backup sources of power, redundant means of electric- and engine-driven pumping, and means to connect portable pumps or standby generation.

How an electrical system is designed and the equipment installed will also affect the reliability of the system. The NEC, in Article 110, covers spacing requirements for electrical equipment. The NEC, in Article 210, covers requirements for receptacle locations in dwelling units, but not for receptacles in industrial areas such as pumping stations (NFPA, 2020a). Receptacle locations for hand tools or welding outlets need to be addressed in the electrical design to ensure proper facility maintenance.

Manufacturer's standards for electrical equipment being installed contain limitations on temperature, humidity, altitude, voltage, interrupting ratings, and clearances.

With the installation of more solid-state electronic controls in pump stations, grounding of the station will affect system reliability. Station grounding is covered by the NEC in Article 250. Also, IEEE's 3003 Standards: Power System Grounding (n.d.-b) is a guide for grounding design. Information on another aspect of system grounding, surge protection, can be found in IEEE Standard C62.62 (IEEE, 2000a).

2.0 VOLTAGE CONSIDERATIONS

The service voltage is chosen after reviewing the facility's loads for connected kilowatts, running kilowatts, and largest individual kilowatts. Contact the utility company about available voltages and requirements for service. The utility company determines what voltage can be supplied to the site based on the requirements of their system. Table 6.1 lists nominal system voltages. The station designer determines the voltage by the power rating of the motors needed to drive the pumping equipment and selects the best standard voltage rated equipment. Table 6.2 presents motor standard voltage ratings.

The power rating ranges for standard motors shown in Table 6.3 represent a manufacturer's typical range for motor construction.

TABLE 6.1 Nominal System Voltages

	Single Phase	Three Phase	
Voltage Class	Three Wire	Three Wire	Four Wire
Low voltage	120/240 V		208 Y/120 V
		240 V	240 /120 V
		480 V	480 Y/277 V
			600 V
Medium voltage		2400 V	
		4160 V	
		4800 V	
		6900 V	
			12 470 Y/7200 V
			13 200 Y/7620 V
		13 800 V	
		24 940 V	13 800 Y/7970 V

TABLE 6.2 Alternating Current Motor Standard Use Voltages

Single Phase	Three Phase
115 V	200 V
230 V	230 V
	460 V
	575 V
	2300 V
	4000 V
	4600 V
	6600 V
	13 200 V

Voltage must be selected with allowances for voltage drop in the circuit wiring. The distance from point-of-service to point-of-use must be considered because of the voltage drop caused by the current draw of the motor and the resistance of the wiring to that current flow. A motor can be operated without damage in a voltage range of plus or minus 10% of the motor's standard voltage rating.

Motor inrush currents caused by starting a motor must be evaluated when selecting the proper voltage. Across-the-line starting of a motor causes

TABLE 6.3 Manufacturer's Standard Power Ranges per Voltage Class

Single Phase	
Voltage Rating	**Power**
115 V	Fractional to 10 hp
230 V	Fractional to 10 hp
Three Phase	
Voltage Rating	**Power**
200/230/460 V	Fractional to 600 hp
2300 V	200 to 4000 hp
4000 or 4600 V	400 to 7000 hp
6600 V	1000 to 12 000 hp
13 200 V	3500 to 25 000 hp

1 hp × 0.746 = 1 kW

locked rotor currents 6 times the motor's full load rated current. These currents are caused by the electromagnetic force necessary to cause the motor shaft to turn and to bring the shaft up to speed. Motor-starting currents can be a significant concern of the electrical utility if they cause flickering on its system as a result of voltage dip at the starting of the motor.

Inrush currents also affect sizing of standby generation equipment because the generator must have the capacity to handle the motor's starting currents. These can be reduced by using reduced voltage starters; wye-delta part-winding starters; solid state, soft-start starters; or adjustable-frequency drives.

The voltage class of equipment being considered needs to be evaluated, as do the training and experience of the maintenance staff. Installing the wiring for medium-voltage equipment requires special skills, and the NEC, in Article 110, discusses space requirements for electrical equipment (NFPA, 2020a). As the equipment voltage class goes up, so does the need for clear space in front of and behind the equipment.

3.0 SUBSTATIONS

Voltage selection may require the installation of a substation to provide the facility on-site transformation of the utility voltage to the equipment's voltage rating. The substation equipment can be owned and furnished by the utility or as part of the facility's equipment. A review of the rate history of the utility for metering on the primary versus the secondary side of the transformer will determine if the cost of the transformation equipment and the assumption of risk should the equipment fail are justified.

The following IEEE standards provide guidelines for the design of a substation and its equipment: (1) Standard 80: Guide for Safety in AC Substation Grounding (2013a) and (2) Standard C37.04: Standard for Ratings and Requirements for AC High-Voltage Circuit Breakers with Rated Maximum Voltage Above 1000 V.

Needs for reliability and redundancy may necessitate a second electrical source at a facility—either a second transformer in a second-source scheme or on-site generation. The risk of electrical system supply failure is lessened with a two-source scheme, be it two transformation points or one source transformation point and generator. A tiebreaker between the two sources is required to add further reliability to the system.

The substation design will need to be coordinated with the utility company, which will need to know the following:

- Where the metering is to be installed,
- What loads are going to be served,

- How many sources of power are required,
- What loads each will serve, and
- What standby capacity each source will be required to have.

The utility may require an easement for the installation of their lines and equipment on the property, and perhaps a one-time charge at the startup of the facility. The utility company typically will recover other costs of providing service, such as transformer losses, in its rate schedule.

4.0 TRANSFORMERS

Guidelines for the manufacture, installation, and testing of distribution, power, and transformers are provided in IEEE Standard C57.12.00: Standard for General Requirements for Liquid-Immersed Distribution, Power, and Regulating Transformers (IEEE, 2021).

As stated in IEEE 3006 Standards (n.d.-b), transformers have a low incidence of failure, but outages are among the longest for a piece of electrical equipment. Utility companies generally have spare transformers in their service yard for most standard pole units and pad-mounted units. A large-station-class unit can take months to replace if it should fail.

Transformers can have an overload capacity above the nameplate rating if forced air cooling is added to the transformer or if the transformer is specified with a higher temperature rise over standard. Transformers come in standard kilovolt ampere (kVA) ratings. Table 6.4 presents single-phase unit ratings and Table 6.5 presents three-phase unit ratings.

Transformers can be divided into two basic groups by their insulating material: dry type or liquid immersed. Both types are designed to handle their full-load rated capacity without additional cooling at an average

TABLE 6.4 Standard Single-Phase Transformer Kilovolt Ampere Ratings

3	75	1250	10 000
5	100	1667	12 500
10	167	2500	16 667
15	250	3333	20 000
25	333	5000	25 000
37.5	500	6667	33 333
50	833	8333	

TABLE 6.5 Standard Three-Phase Transformer Kilovolt Ampere Ratings

15	300	3750	25 000
30	500	5000	30 000
45	750	7500	37 500
75	1000	10 000	50 000
112.5	1500	12 000	60 000
150	2000	15 000	75 000
225	2500	20 000	100 000

ambient temperature of 30 °C (85 °F) for a 24-hour period, with the maximum temperature not exceeding 40 °C (105 °F).

For a dry-type transformer, the temperature of the cooling air when the unit is operating should not drop below −30 °C (−22 °F), and the altitude of the unit should not exceed 1000 m (3300 ft) without derating the unit.

For a liquid-immersed, air-cooled transformer operating without additional cooling, the surface temperature of the liquid should not fall below −20 °C (−4 °F). Similarly, the altitude should not exceed 1000 m (3300 ft) without derating the unit.

Atypical transformer service conditions need to be addressed in the design and layout of the equipment. Such conditions occur when ambient temperatures rise or fall or through exposure to damaging fumes, vapors, abrasive dusts, salt spray, excessive moisture or dripping water, abnormal vibration, shock, tilting, and seismic conditions.

Installation requirements for transformers are covered in NEC Article 450, American National Standards Institute (ANSI)/IEEE C57.12.01 for dry-type transformers and ANSI/IEEE C57.12.00 for liquid-immersed transformers (IEEE, 2021).

Transformer impedance is a large factor in short-circuit studies that controls the maximum amount of fault current let through to the secondary side. The higher the impedance of the transformer, the lower the fault current will be on the secondary side of the transformer.

5.0 SWITCHGEAR

Switchgear provides reliability in the facility design. By splitting the main bus of the switchgear with a tie-interrupting device, the facility loads can be divided between halves of the equipment, allowing isolation in case of equipment failure of an upstream device. The tie enables the halves to be connected, and this arrangement, as illustrated in Figure 6.1, is called a

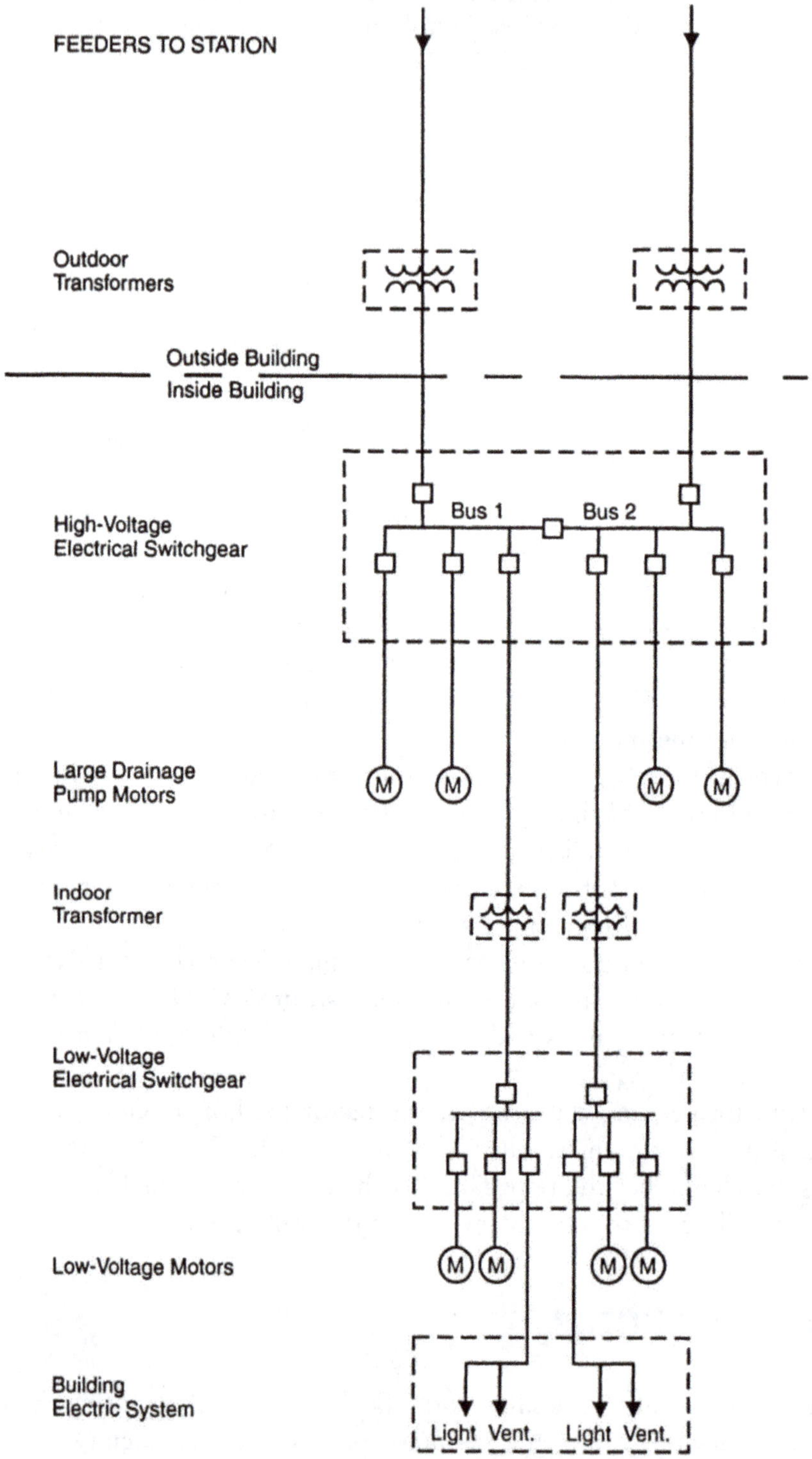

FIGURE 6.1 One-Line Diagram for a Stormwater Pumping Station

selective system. A switchgear located on the primary of a transformer is called a *primary selective system*. Having the tie on the secondary side of the transformer provides a secondary selective system.

Requirements for the design, installation, and testing of switchgear assemblies are discussed in IEEE Standard C37.04: Standard for Ratings and Requirements for AC High-Voltage Circuit Breakers with Rated Maximum Voltage Above 1000 V—Corrigendum 1 (IEEE, 2018).

There are two types of switchgear assemblies: open and metal enclosed. In open switchgear, the main bus bar, interconnecting wiring, interrupting device, associated controls, and metering are not enclosed in a supporting structure. In metal-enclosed switchgear, these items are enclosed on all sides and the top by a metal supporting structure.

Requirements for working space around this equipment are based on the equipment's use voltage (NFPA, 2020a). Installation requirements of the manufacturer cover working space required for the removal of interrupting devices for the switchgear assembly.

Table 6.6 covers standard continuous current rating for buses of metal-enclosed switchgear.

The interrupting devices that make up the main component of this equipment are named for the insulating material used in their design. In medium-voltage equipment, the interrupting of the circuit is done in a vacuum, in air, with a gas (SF6), or with fuses. The selection of interrupting device used at medium voltage is based on the fault-interrupting capacity of the device, the speed of fault interruption required, the space required to house the device, and the cost. Coordination among interrupting devices used to be a concern, especially between systems with fuses and relaying, but electronic fuses have solved the problem of coordinating a fused switch with a relay-based protection system.

TABLE 6.6 Manufacturer's Standard Bus Ratings by Equipment Voltage Class

To 600 V	From 600 to 13 800 V
600 A	—
1200 A	1200 A
1600 A	—
2000 A	2000 A
3000 A	3000 A
4000 A	—

The reliability of an electrical system is affected by how well its interrupting devices can be coordinated. Proper coordination of the protective devices is paramount during a fault condition on an electrical system. The device closest to the fault needs to be opened to clear a fault from the system while an upstream device is set to allow passage of the fault current necessary for operation of the downstream device without tripping itself. Relay schemes, fuse curve coordination, and operating speed of interrupting devices play important roles in this process. The separation between current characteristic curves of a coordination study at the maximum let-through current should be greater than either the operating time of the interrupting device that is to open on fault or the melting time in case of a fuse.

The advent of microprocessor-based metering and relay equipment allows for a more selective coordination of system components and for interfacing with an industrial network so that settings and load information can be displayed at remote locations. For large motors, these devices have provided for custom protection of motors in one relay.

6.0 CONDUIT AND WIRE

The NEC defines permitted uses for each type of conduit cable support system, the installation requirements for supports, and grounding concerns (NFPA, 2020a). Based on NEC data, Table 6.7 catalogs the areas where different types of conduits can be used.

The tables in Article 310 of the NEC, which provide the ampere capacity rating for wire and cable up to 15 000 V, are based on the metal used to conduct the electricity, either aluminum or copper. Notes following the tables should be consulted for wire and cable installation derating requirements. The NEC provides tables for the conduit fill requirements for 600-V

TABLE 6.7 Areas for Permissible Use of Conduit

Classification	Exposed	Direct Burial	Hazardous Areas
Rigid metal conduit	Yes	Yes	Yes
Intermediate metal conduit	Yes	Yes	Yes
Electrical metallic tubing	Yes	Yes	As noted in NEC[a]
Rigid nonmetallic conduit	Yes	Yes	As noted in NEC
Flexible metal conduit	As noted in NEC	No	As noted in NEC
Electrical nonmetallic tubing	As noted in NEC	No	No

[a]NEC = National Electrical Code

rated cables. For cables rated above 600 V, the manufacturer's standards for cable construction will need to be consulted for cable dimensions and uses.

Guidelines for specifying electric submersible pump cable are provided in IEEE Standards 1018 (2013b) and 1019 (2013c). For wiring practice in hazardous areas, IEEE Standard 576 (2000b) and the NEC can be reviewed.

Sizing requirements of circuits for motor feeder and service conductors and the conduit systems that support them are covered in the articles of the NEC. Article 430 discusses requirements for sizing the feeder cables and disconnect for a motor circuit, and tables provide listings of motor full-load currents, which are useful when the nameplate data of the motor are not known. The listed motor ampacity is multiplied by 125% to calculate the design load for sizing feeders and disconnect for the motor, and listed motor ampacity is multiplied by 150% to calculate the full-load ampacity of the drive. The drive-calculated, full-load ampacity is then multiplied by 135% to calculate the design load for sizing feeders and disconnect for the drive.

Not covered in the NEC are sidewall pressure limitation on cables during installation, pulling tension limitations of cables during installation, and the fault-withstand rating of wires and cables (i.e., the amount of current a cable can carry over time without damage to its insulation). The cable fault withstand must be considered when sizing feeder circuits for proper coordination during fault conditions for downstream devices. Sidewall pressure and pulling tensions should be addressed in the design of the conduit system.

When designing a medium-voltage system, the designer needs to be familiar with special requirements and considerations of cables.

Cable testing is recommended to verify the reliability of the electrical system after it is installed and also as a means of periodically determining the condition of the system. Testing of cable insulation with a meggar or by direct current voltage provides a basis upon which to compare future cable tests.

7.0 LIGHTNING PROTECTION

Requirements for the design of a lightning protection system can be found in NFPA Standard 780 (NFPA, 2020b).

Lightning is a spark averaging 25 000 A and 30 000 V that discharges between the ground and a cloud in 50 microseconds across a path from a few thousand feet up to several miles. The discharge is made up of multiple pulses and generates heat up to 16 500 °C (30 000 °F) in a matter of microseconds. The heat generated by a lightning stroke can cause significant damage to structures by converting moisture in building materials to high-pressure steam that can cause concrete and mortar to crack and lumber and fiberglass to shatter. The heat can also vaporize, weld, and melt electrical and electronic components.

Lightning hazards include side flashes caused when adjacent objects are at different potentials because of the sudden and unequal rise in ground potential. Side flashes balance the potential between objects.

A structure's lightning protection system needs to intercept the lightning stroke, divide the current, and conduct it safely to the ground past the structure and its contents and occupants. The lightning protection system requires a low-resistance ground system that provides equal ground potential for all parts of the structure and its contents and occupants.

Another form of lightning protection is surge protection. Most lightning strikes are cloud to cloud, but this type of lightning induces a voltage to electrical and electronic components. Standards for components designed to handle induced currents are found in IEEE Standard C62.62 (2000a).

Surge protection should be installed on telecommunication systems, security systems, fire alarm systems, motor control centers, switchboards, switchgear, computers, standby generators, surveillance systems, instrumentation, and all systems that communicate or are connected from the facility by wire.

8.0 STANDBY POWER

Guidelines for the installation of standby power can be found in IEEE 3005 Standards: Energy & Standby Power Systems (IEEE, n.d.-a). The need and requirements for a second source of power to a pump station are dictated by the reliability of the primary power source and the requirements of local jurisdictions.

The standby source of power could be a second utility line if arrangements can be worked out with the utility serving the facility. The cost of providing a second source, which can be charged to the facility owner, may dictate installation of an on-site source of standby power (see Figure 6.2) as a more economical approach.

Engine-driven pumps are another possible secondary source of power. Space must be provided in the design for any fuel storage facility. Standby power should be provided for any fuel pumping needed to transfer fuel to the engine from the fuel storage facilities.

Standby power can also come from an engine and generator set. Sizing of the unit depends on the loads that will be applied to the unit, the sequence in which the loads will be applied, and the type of loads being applied. Depending on the loads required to operate off of the generator, multiple generators may be required to operate in parallel. A generator sizing

FIGURE 6.2 Standby Power Generator

analysis will be required to be performed to determine the number and sizes of generators required. When conducting the generator sizing analysis, the maximum voltage dip desired and ambient temperature the generator will operate in need to be specified. Typical maximum voltage dip is 15%.

If variable-frequency drives (VFDs) are included as part of the loads served by the generator, then the supplier of the generation equipment needs to be informed of these loads. Variable-frequency drives cause harmonic distortion, which can affect the generator's voltage regulator if it does not have three-phase compensation. The generator supplier may recommend an oversized generator to compensate for the effects of overheating in the generator wiring caused by harmonic distortion.

Recommendations of the engine manufacturer should be followed for intake air, exhaust, and cooling requirements. When an engine-driven pump or engine and generator set is installed, the designer must consider maintenance requirements of the engine. Space and circuits need to be provided for starting batteries, battery charger, engine heater, remote radiator, day tank, fuel level probes, and other engine ancillary equipment.

Regarding the transfer scheme in the event of power failure, the time the facility can be without power must be determined. A delay period for the transfer scheme can be specified to allow the station to ride through temporary power outages. If the generator is sized to handle only part of the loads, the transfer scheme should contain contacts to lock out loads that overload the generator. A time delay period can be specified to delay operation of the transfer scheme when operating on the standby source, and this period starts when the utility source becomes available. It maintains operation of the supply voltage from the standby source to verify that the utility source voltage is stable before the transfer takes effect.

In transferring motor loads from the standby power source back to the normal utility source, the transfer scheme can handle motor load transfer with in-phase transfer between sources, a time delay in a neutral position between the two sources during transfer, or by shutting off the motors before transfer between sources. Coordination with the utility should take place for any closed transition scenarios when switching from generator power back to utility power.

9.0 ADJUSTABLE-SPEED DRIVES

Alternating current VFDs are the most common type of adjustable-speed drives used in wastewater and stormwater applications. They allow pump-driven alternating current motors to be operated at adjustable speed, resulting in adjustable flow.

The option of adjusting the speed of alternating current motors promotes their use over direct current motors. More specifically, alternating current squirrel cage induction motors are simple, capable of high speed, require little maintenance, and typically perform well under such environmental conditions as dirty air and explosive atmospheres.

For varying torque loads such as centrifugal pumps, VFDs are capable of controlling speed over a 10:1 range. Centrifugal pump speed ranges typically are limited by the pump remaining within its operating envelope of the system head curve. Because this limitation is almost always less than 10:1, VFDs rarely limit the speed range of centrifugal pumps.

9.1 Matching Variable-Frequency Drives and Motors

Variable-frequency drive systems require proper matching of the motor and drive, and mismatches sometimes occur. Torque requirements at any point in the operating speed range play a role, and a common mistake is to use the nameplate power rating to match drive and motor. Because different motor types have different power factors, full load amperage draw may differ even though the power rating is the same. Full load amperage draw should be used in matching motor and drive. Variable-frequency drives are rated not on horsepower but on the amperage of the load they are driving.

Several methods have been used in the design documents to promote the matching of motor and drive. In one, the pump supplier is assigned complete responsibility for the match and purchases the VFD from a vendor. A drawback with this approach is the possibility of projects having different types of pumps supplied by multiple pump suppliers. Obtaining VFDs of only one manufacturer for the entire project can prove difficult. For stations where one type of pump is specified, this is less of a problem.

In another method, the electrical subcontractor supplies the VFDs for the entire project and mounts them together with the motor control centers for constant-speed pumps. The electrical subcontractor assumes some responsibility for the match but does not purchase the pumps. This promotes installations that have only one VFD manufacturer and typically makes for a well-coordinated motor control center. However, because the electrical subcontractor does not purchase the pumps, the responsibility of correcting the problem is less clear than in the first method if a mismatch occurs.

During the construction phase of the project, regardless of who provides the VFD and motor, a letter of compatibility shall be issued from both the motor and VFD manufacturers stating that both the motor and VFD are compatible with each other. Also, to alleviate any issues with the motor, it is recommended that any motors driven by a VFD be inverter duty rated and the motor bearings be insulated to alleviate circulating currents caused by the pulse width modulated signal of the VFD from damaging the motor bearings.

9.2 Variable-Frequency Drive Interfaces

Variable-frequency drives are required to interface with the electrical system and the instrumentation and control system. Proper coordination of control circuits and speed signals with these systems is essential.

It is recommended that control circuits be 115 V–ac, which is standard or an available option on most drives other than some smaller VFDs. For these VFDs, the control circuit is low alternating current or direct current voltage (e.g., 12 V–dc). Low-voltage control circuits are prone to outside interference such as from voltage induced from nearby 460-V, three-phase alternating current power.

Often, VFDs are controlled by an external analog signal, which must be compatible with the process control system. This signal typically will be a 4- to 20-mA speed input as opposed to a varying voltage signal such as 1- to 5-V direct current. Similarly, the speed output signal must be compatible with the process control system and also typically will be 4 to 20 mA and isolated, if possible. The standard interface signals are varying voltage such as 1 to 5 V direct current, especially for some smaller VFDs. Thus, if not properly specified, field problems can result.

9.3 Some Problems With Variable-Frequency Drives

Many VFDs are typically rated for service between 0 and 40 °C (32 and 104 °F), with a noncondensing relative humidity of up to 95%. They typically are installed in a non-air-conditioned electrical room. In years past, the electrical room was air conditioned to provide the VFD solid-state components with a mild environment; this practice may still be appropriate for selected applications. Variable-frequency drive manufacturers offer VFDs that are rated to operate continuously in a 50 °C (122 °F) ambient temperature.

If a bubbler tube is used for level control, avoid mounting the bubbler tube pressure transmitter in the VFD enclosure or it might introduce corrosive gases such as hydrogen sulfide, damaging the VFD electronics.

The solution to harmonics and notching problems is to reduce the total harmonic distortion of the incoming power supply. Typical solutions include adding filters and reactor circuits to the VFD. Reducing harmonics involves not only the drive itself, but how it interacts with the incoming power, and some practitioners recommend that design documents emphasize this interaction. For instance, the VFD drive vendor might receive more information on the electrical system design than is now commonly issued.

A harmonic analysis should be conducted for any pump station that has VFDs installed. The analysis should be conducted before any electrical equipment is ordered. The applicable standard for harmonic control is IEEE (2014) Standard 519, which places limits on such items as total harmonic distortion.

10.0 POWER FACTOR

Depending on the local utility requirements, power factor correction capacitors may be required to alleviate a power factor penalty imposed by the local utility for having a low power factor. For instance, the local utility may have a minimum power factor requirement of 95%. A power factor less than 95% will cause the facility to pay a monthly penalty for having less than a 95% power factor.

Power factor correction capacitors are installed to improve the power factor. The capacitors can be installed on individual motors and switched on when the motor is energized or installed in a switched bank. When the capacitors are connected to individual motors, coordination with the motor manufacturer needs to take place to determine the maximum amount of kVARS that can be connected to the motor without overexciting the motor. Typically, the maximum the power factor can be corrected to on an individual motor is 95% without overexciting the motor. Motors that are driven by adjustable speed drives do not need power factor correction connected to them because, internally, the drive corrects the power factor to 95% or better. Capacitors can be medium voltage and low voltage.

A switched capacitor is typically comprised of multiple steps of capacitors that are brought online to correct the power factor to a certain setpoint value. The stepped bank has a power factor controller that monitors the power factor and turns on/off the steps, as required, to achieve the power factor setpoint desired. Stepped banks are typically connected to the main incoming switchgear or other locations in the electrical distribution system.

An analysis is recommended to determine the optimum location and type of capacitor bank, individual or switched, to use to correct the power factor. Also, a harmonic analysis and capacitor back-to-back switching analysis should be performed during the design phase.

11.0 ARC FLASH HAZARD ANALYSIS

As part of the electrical installation, an arc flash hazard analysis should be conducted for the electrical system. An arc flash hazard analysis is a study of the facility's power system to determine the incident energy available at specific electrical equipment that operations and maintenance personnel would be exposed to while operating the electrical equipment. The results of the analysis are a determination and labeling of the personal protective equipment required of the electrical equipment, suitability of the equipment for service as installed, and if there are any mis-coordination issues involved as the system is currently installed. The goal is to provide the calculated arc flash hazard of the particular electrical equipment as installed and to provide labels for the use of maintenance personnel in determining the proper protective equipment to wear. The arc flash hazard analysis shall be conducted in accordance with NFPA 70E, Standard for Electrical Safety in the Workplace (NFPA, 2021).

12.0 REFERENCES

Institute of Electrical and Electronic Engineers. (n.d.-a). IEEE 3005 Standards: Energy & Standby Power Systems.

Institute of Electrical and Electronic Engineers. (n.d.-b). IEEE 3006 Standards: Power Systems Reliability.

Institute of Electrical and Electronic Engineers. (2000a). IEEE C62.62: IEEE Standard Test Specifications for Surge Protective Devices for Low Voltage AC Power Circuits.

Institute of Electrical and Electronic Engineers. (2000b). IEEE 576: IEEE Recommended Practice for Installation, Termination, and Testing of Insulated Power Cable as Used in Industrial and Commercial Applications.

Institute of Electrical and Electronic Engineers. (2013a). IEEE 80: IEEE Guide for Safety in AC Substation Grounding.

Institute of Electrical and Electronic Engineers. (2013b). IEEE 1018: IEEE Recommended Practice for Specifying Electric Submersible Pump Cable—Ethylene-Propylene Rubber Insulation.

Institute of Electrical and Electronic Engineers. (2013c). IEEE 1019: IEEE Recommended Practice for Specifying Electric Submersible Pump Cable–Polypropylene Insulation.

Institute of Electrical and Electronic Engineers. (2014). IEEE 519: IEEE Recommended Practice and Requirements for Harmonic Control in Electric Power Systems.

Institute of Electrical and Electronic Engineers. (2018). IEEE C37.04: IEEE Standard for Ratings and Requirements for AC High-Voltage Circuit Breakers with Rated Maximum Voltage Above 1000 V—Corrigendum 1.

Institute of Electrical and Electronic Engineers. (2021). IEEE C57.12.00: IEEE Standard for General Requirements for Liquid-Immersed Distribution, Power, and Regulating Transformers.

National Fire Protection Association. (2020a). NFPA 70, National Electrical Safety Code.

National Fire Protection Association. (2020b). NFPA 780, Standard for the Installation of Lightning Protection Systems.

National Fire Protection Association. (2020c). NFPA 820, Standard for Fire Protection in Wastewater Treatment and Collection Facilities.

National Fire Protection Association. (2021). NFPA 70E, Standard for Electrical Safety in the Workplace.

13.0 SUGGESTED READINGS

Institute of Electrical and Electronic Engineers. (n.d.-a). IEEE 3001 Standards: Power Systems Design.

Institute of Electrical and Electronic Engineers. (n.d.-b). IEEE 3002 Standards: Power Systems Analysis.

Institute of Electrical and Electronic Engineers. (n.d.-c). IEEE 3003 Standards: Power Systems Grounding.

Institute of Electrical and Electronic Engineers. (n.d.-d). IEEE 3004 Standards: Protection & Coordination.

Institute of Electrical and Electronic Engineers. (2017). IEEE C2: National Electrical Safety Code.

Institute of Electrical and Electronic Engineers. (2020). IEEE C5712.01: IEEE Standard for General Requirements for Dry-Type Distribution and Power Transformers.

7

Instrumentation and Control Systems

Control technology for wastewater and stormwater pumping stations has advanced greatly over the last few decades, as a result of the following:

- The microprocessor, which now pervades control applications. Programmable controllers, digital telemetry systems, and transmitters are all microprocessor based.
- Cell phones and tablets that allow an operator to be in the field and remotely access information from multiple sites.
- Advancements in communications that allow for a larger amount of data to be gathered and analyzed.

To the pumping station designer, these advances offer increased reliability, improved ways to meet design objectives and alternate equipment, a reduction in labor, and valuable data.

Relevant design objectives include proper functioning of the pumping station and proper management of the overall wastewater or stormwater pumping system. Improved ways of alternating equipment can extend equipment life, and a reliable pumping system reduces overflows and surcharges. A reliable telemetry system often reduces labor and operation costs. Data from catchment areas can identify infiltration and inflow problems. Flow data assist in designing for expansions. Data characterizing diurnal patterns can identify frequently operating equipment, which can be incorporated into a maintenance program.

1.0 PROCESS AND INSTRUMENTATION DIAGRAMS

The use of process and instrumentation diagrams (P&IDs) and process flow diagrams is suggested in the design of wastewater and stormwater pumping stations because they provide an efficient means of communicating design concepts. Use of P&IDs is recommended even for simple stations, although the designer may choose not to include them as part of the design documents.

Figure 7.1 presents an example P&ID of a basic pumping station. Symbols of P&IDs should conform to International Society of Automation Standards and Practices, specifically American National Standards Institute/International Society of Automation (ISA) 5.1—Instrumentation Symbols and Identification (ISA, 2009).

Within their simplicity lies the value of P&IDs. The design of a pumping station is multidisciplinary, and P&IDs promote communication between process mechanical, instrumentation and control (I&C), and electrical designers. They promote proper piping and process concepts and the creation of an accurate instrument list and assist in establishing control concepts and subsequent development of control diagrams. Accurate electrical drawings with detailed conduit and conductor runs have as their origin correct P&IDs. With P&IDs, pumping station operations personnel can input their ideas to the design.

During design, P&IDs are not static and reflect the changes brought about by the design team itself, quality assurance reviewers, client preferences, and value engineering efforts. However, the design effort is typically

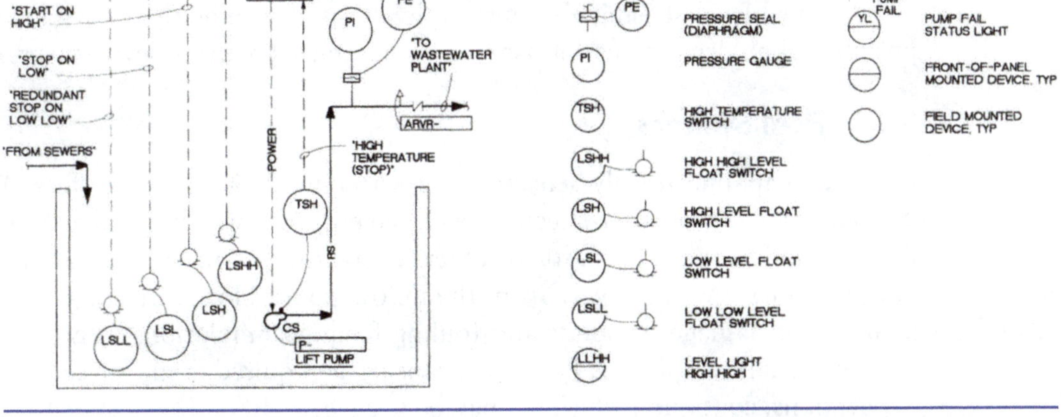

FIGURE 7.1 Example Process and Instrumentation Diagram of a Basic Pumping Station

improved if the P&IDs are "frozen" early in the design, with only minor changes made as the design progresses.

2.0 LEVEL MEASUREMENT

Early in the design, the required operating level range should be determined because it affects vertical dimension of the wet well and hence the structural design. Often overlooked is the fact that enough vertical dimension should be reserved to sense abnormal low and high levels (besides normal high and low operating points). Also suggested is to sense these abnormal levels by primary elements independent of the elements used to sense normal operating levels.

The abnormal low level should be at least equal to the center line elevation of the impeller. The conservative choice is to locate the abnormal low point equal to the volute top to reduce air entry into the volute. In the case of submersible pumps, the submersible motor may add another complication. The conservative choice is to locate the abnormal low point above the top of the submersible motor for cooling, increasing the required vertical dimension even more. However, most submersible pumps can operate properly with their motors only partially submerged. If the designer wants to reduce the vertical dimension or operate the submersible pump at times with its motor not submerged, the pump manufacturer should be contacted for a recommendation.

The normal low operating level is typically set at least 100 to 150 mm (4 to 6 in.) above the abnormal operating low level, though the normal low operating level should be confirmed with the pump vendor. Based on wet well geometry and piping layout, vendors sometimes recommend a normal low operating level to ensure proper operation (e.g., reducing vortexing effects to meet minimum net positive suction head requirements).

Wet well level sensing elements typically used include float switches, pneumatic bubbler systems, ultrasonic sensors, and submersible level sensors.

2.1 Float Switches

Although capable of only sensing fixed setpoints, float switches (floats) typically are the most cost-effective level element. To resist chemical action, floats should be made of plastic or other resistant material. They should be put in a quiescent area away from the inflow point. Floats are subject to jamming or damage by solids and fouling from accumulations of grease, fats, oil, scum, and debris. Periodic cleansing is required, and, for certain applications, continuous flushing may be needed.

Floats can be mounted either to a vertical pipe (typically 25 mm [1 in.] in diameter) or to a suspended cable furnished with a weight at its bottom. Vertical pipe mounting is suggested for most applications because the setpoints are reproducible and it requires less space in plan view than suspended cable mounting, which requires more area because the float can move 360 deg in plan view on a suspended cable. Also, debris can cause the suspended cable to rise, resulting in malfunction. Suspended cables do allow the changing of setpoints to be made easier than with the vertical pipe. With either mounting, float setpoints may be as close as 100 mm (4 in.) apart.

To prevent tangling of close setpoints, floats should be staggered on each vertical pipe or mounted on different pipes. In a staggered arrangement, in plan view, the floats are not mounted on top of each other, but rather form angles with each other.

As with all primary elements, level elements must be approved for use in their surrounding atmosphere. Specific care should be given to floats because they are located in the wet well itself. Wastewater wet wells are often classified as Class I, Division 1, hazardous locations. Accordingly, float circuits must be provided with intrinsically safe relays.

If the wet well is classified as hazardous, then conduits and conductors must meet required codes for use in hazardous atmospheres (see Chapters 3 and 6 for more details).

2.2 Bubbler Systems

The pneumatic bubbler system can either monitor discrete level setpoints with pressure switches or continuously measure wet well level with a pressure transmitter. The bubbler system also requires a compressor, receiver, ancillaries, and a bubbler tube immersed in the wet well. Ancillaries include such items as a pressure-regulating filter, constant differential pressure relay, a blowdown valve, and isolation valves.

Except for the bubbler tube itself, these components are available in an integral control panel, which is useful in isolated pumping station cells. Where the pumping station is adjacent to a facility such as a water resource recovery facility (WRRF), the bubbler system is more custom designed, often making use of existing high-pressure air systems. The receiver typically is furnished with a low-pressure switch, which monitors compressor malfunction. As with floats, the tube is subject to fouling. To clear the tube, provisions typically are made to subject it to a high-pressure air purge, which can be done manually or automatically at periodic intervals. Sometimes provisions are made to physically rod the tube.

Bubblers generally are reliable and cost effective. However, especially because of the number of fittings, some systems are subject to leaks, leading to erratic results. During cold winter months, some systems perform erratically because of freezing of condensed moisture within the tubing.

In corrosive environments, the working life of bubbler systems has been reduced by corrosion of system components. Accordingly, component materials should be as noncorrosive as possible. Bubbler systems located indoors can be installed in the same room as the electrical switchgear, though corrosion problems with electrical equipment have been experienced in some locations because of corrosive gases introduced by bubbler tube leaks. This has led some practitioners to avoid locating bubbler systems with electrical switchgear.

A bubbler tube 13 mm (0.5 in.) in diameter with its end approximately 150 mm (6 in.) from the wet well bottom has proven adequate. Some commercially available bubblers deliver an airflow of approximately 0.2 L/s (0.4 scfm), though some practitioners prefer a higher flow of approximately 0.5 L/s (1 scfm) (Bergs, 1993).

Pressure switch designs should be critically reviewed because the commercially available ranges of pressure switches are limited. In certain wastewater and stormwater applications, suitable pressure switches are only available in the low end of the ranges. Switch deadband is sometimes large enough at some low ends to render the switch nonreproducible and useless (Whitenack, 1992).

Figure 7. 2 shows a typical bubbler system within a control panel. This system happens to use pressure switches, though the compressor is not integral to the panel. Note the high-pressure purge bypass around the normal operation piping. Figure 7.3 depicts the top of a bubbler tube furnished with a plug, which can be removed to allow rodding of the tube.

2.3 Ultrasonic Sensors

Ultrasonic level sensors are nonintrusive in nature, which makes them less subject to fouling and debris. Particularly in stormwater pumping applications, they are often the element of choice because they are typically not affected by debris.

On occasion, ultrasonic level sensors are affected by large floating materials. They are confused by foam, vortices, turbulence, and wall echoes. Where the ultrasonic transducer does not include its own reference target for temperature or humidity correction, thermistor temperature monitoring and measurement correction have been incorporated to some transducer designs. The thermistors can fail and render the transit time correction algorithms ineffective. While the sensor is nonintrusive (i.e., not immersed in the wet well liquid), it must still be mounted in the well, exposed to the ambient atmosphere.

The sensor must be mounted far enough away from walls to prevent wall echo interference. Mounting should also allow sufficient headroom

FIGURE 7.2 Bubbler System Internals

for the ultrasonic emission dead zone (typically 460 to 610 mm [18 to 24 in.] above the absolute maximum liquid elevation level). Sometimes wet well bottoms are grouted to form sharp angles to prevent solids and debris from gathering in corners. If not accounted for, such grouted surfaces may interfere with the desired range of the sensor. A wastewater wet well with a closed access hatch and no grating can be considered a closed tank. Some, but not all, ultrasonic level meters require that the bottom of their megaphones be at least 50 mm (2 in.) above the vent pipe of the wet well to ensure safe operation if the wet well is classified as Class I, Division 1, and is considered a closed tank.

2.4 Radar Level Sensors

Because of advancements in technology, radar level sensors are becoming more common and more accurate than ultrasonic level sensors. Radar level sensors have an accuracy of +/−2 mm up to 40 ft and have the benefit of being able to measure solids and be used in the same applications as

FIGURE 7.3 Bubbler Tube Top

ultrasonic level sensors. In addition, radar is not temperature dependent like ultrasonic is.

Radar level sensors measure level by emitting microwaves and use time-of-flight technology to measure the distance based on the time it takes for the microwave signal to reflect back. There are two types of radar level sensors: non-contact and guided. For most applications, a non-contact radar level sensor can be used.

When specifying radar, the designer will need to select the frequency band and antenna geometry that will be used because both affect the performance of the sensor. Three different frequency bands are typically available, with the K-band (ca. 26 GHz) being the most common. Selection of the frequency band is dependent on what you are measuring through. The antenna geometry refers to the shape of the beam, but also the shape of the antenna. Additional consideration should be given if condensation is expected. In those applications, the designer should look for a sensor that is not flat but is shaped so condensation (i.e., water droplets) does not build up.

It is possible to flange-mount the radar level sensor or suspend it over the media and make it accessible from a nearby hatch. Similar to ultrasonic, the designer needs to take into consideration the placement of the sensor to avoid interferences that may be in the path of the beam angle, such as guide rails, pumps, and piping. Several manufacturers allow for false echo dedication that allow the instrument to tune out static objects that interfere with the beam angle.

2.5 Submersible Level Sensors

Submersible level sensors are submerged near the bottom of the wet well and measure level by sensing hydrostatic head pressure.

The sensor itself is a polytetrafluoroethylene-faced diaphragm connected by cables to a transmitter housing. The diaphragm is part of a sealed, oil-filled system. A variable-capacitance transducer converts the pressure change to an electronic signal. As with any pressure-based system that measures level, specific gravity of the measured fluid must remain constant, which is essentially the case for wastewater and stormwater.

A sealed breathing system is a key part of the system and can accommodate cable lengths of up to approximately 24 m (80 ft). The breathing system protects the internals of the submerged sensor, allows for expansion and contraction, and nulls the effects of barometric changes.

These units, available either loop powered or externally powered, are not intrinsically safe, but can be made so using intrinsically safe barriers and other arrangements.

The popularity of these devices is increasing, and practitioners generally have been pleased with this type of system. Because they are submersed, they have particular value where freezing is a concern.

These units can be mounted either on a vertical pipe or by means of a suspended cable. As with float switches, the conservative choice is to pipe mount.

3.0 FLOW MEASUREMENT

Flow meters are installed on most pumping stations. In smaller collection systems, flow meters might not be used. Flow meters typically will be installed on larger pumping stations that discharge directly to WRRFs. For wastewater or stormwater systems in which flow is managed systemwide, flow meters are prevalent.

An accurate flow meter facilitates pump field tests, which are difficult to perform precisely even under the best of circumstances. The acquired flow data also assist in systemwide planning for future expansion. Instantaneous flow data aid in the operation of wastewater or stormwater systems where diversion and storage are practiced.

Flow can be measured either in a closed pipe or an open channel. Closed-pipe flow measurement is more common in wastewater and stormwater pumping stations. Open channel flow measurement typically is reserved for those pumping stations with screening facilities that require an open channel. Commonly used closed-pipe flow meters include magnetic flow meters, ultrasonic flow meters, and differential producers. Open channel flow measuring systems include flumes and nozzles and the "secondaries" that measure their levels and convert them to flow. Commonly used "secondaries" include ultrasonic and admittance sensors.

All closed pipe flow meters will not measure accurately if a pipe is partially full, which can occur in horizontal runs under low or no flow conditions. If such flow conditions are expected, make provisions to reject the flow measurements by monitoring the process fluid level in the pipe. This "full/not full" switch can be used to discriminate and select flow information obtained only during suitable measuring periods.

Early in the design, inline meters should be selected and sized. Such inline meters include magnetic flow meters, flow tubes, and classic Venturi meters. By themselves, these meters typically do not produce a significant (non-recovered) head loss. However, to measure accurately, these meters may need to be less than line size, and the fittings required can produce a significant head loss. Although more of a problem with gravity systems, this head loss can affect pump selection.

For both open and closed channel flow meters, a certain number of straight upstream and downstream diameters are required for accurate flow measurement. This number depends on both the type of meter and the type of upstream or downstream obstruction. Generally, upstream effects are more pronounced, requiring longer straight upstream runs than downstream runs.

There are many pumping stations in which flow is consistently measured accurately, but also many stations where flow measurement is inaccurate. Accurate flow measurement probably is less common than accurate level or pressure measurement. Besides following guidelines found in this and other books, manufacturers' recommendations found in their operations manuals and metering guides should be followed.

The design objective of flow measurement is to provide accurate and reproducible data. Occasionally, only reasonably reproducible and not overly accurate flow data are required. Thus, recommended flow meter installation practices sometimes are violated. A common example is installing flow meters to existing piping at less than recommended straight upstream diameters. In these instances, owners should be apprised of risks such as unacceptable reproducibility before installing flow meters.

Published accuracy of flow meters typically is done as either a percentage of "rate" (accuracy as compared to the actual flowrate) or a percentage of "span" (accuracy as compared to the full-range flowrate). Percent of "rate" values are clearcut. Percent of "span" values are less so because there may be unacceptable inaccuracies at low-end flows.

3.1 Magnetic Flow Meters

Magnetic flow meters are well suited for stormwater and wastewater applications because they are chemically resistant if properly specified and are little affected by sediments and debris. Their accuracy typically is ±1% of rate over a flow range of 10:1. Suitable liner material for wastewater is either polyurethane or polytetrafluoroethylene. In most cases, polyurethane is as suitable as polytetrafluoroethylene and is the liner of choice because it is less costly. Though 316 stainless steel electrodes typically are used, there are other satisfactory electrode materials.

There are two basic types of magnetic flow meters, alternating current and pulsed direct current, which refer to how the magnetic field of the meter is excited. The pulsed direct current type is predominant in wastewater and stormwater pumping applications. Although its field is excited by pulsed direct current, supply power for this type of meter typically is 120 V–ac.

Magnetic flow meters typically require at least five straight upstream and two or three straight downstream pipe diameters. Models are available with even shorter requirements.

Exceptions to the five straight upstream diameter rule are the distances from control valves and from pump discharge, for which the meter may require 10 straight upstream pipe diameters. The 10 straight upstream rule is necessary to ensure high accuracy (for example, 0.5% of rate, of which some modern microprocessor-based magnetic flow meters are capable). Often, 10 straight upstream pipe diameter is impractical to achieve, and high accuracy is not needed in stormwater and wastewater applications. The meter manufacturer should be contacted when a shorter straight upstream run is considered.

Where magnetic flow meters are less than line size, some manufacturers permit concentric reducers to be placed immediately on both sides of the meter.

Magnetic flow meters are available with transmitters mounted either on the meter body itself or remotely from it. Remote-mounted transmitters typically are located near the meter, though they can be located farther away. Remote-mounted transmitters reduce possible damage from hydraulic transients and slamming check valves.

Besides watertight ratings, magnetic flow meters are available in options for accidental (up to 48 hours) and even continuous submergence use (both of up to approximately 9 m [30 ft]). Some manufacturers also place an upper time limit on their continuous-submergence magnetic flow meters. The designer should be conservative in option selection because moisture in the electronics can render the meter useless.

Magnetic flow meters have several standard or optional features that may be of benefit. Some have a zero-return feature that can be interlocked with either a pump or a device measuring pipe level. If the pumps stop or if a pipe is partially empty, the meter reading will be zero. Low-flow cutoff set typically at 1% or 2% will suspend meter operation. Sloshing within the pipe at actual zero flow may produce a false flow reading for this switch. As an option, magnetic flow meters are capable of reading bidirectional flow, which might be of use in some applications.

Entrapped air will produce erroneously high readings. In many pumping stations, the horizontal line in which the magnetic flow meter is placed is a local high spot. Such installations are potential sources of error because of entrapped air, and manual and automatic air release valves should be used to release air. Preferred practice is to lower the meter below the horizontal line, or to slope the horizontal line piping so downstream piping is raised enough to guarantee a full magnetic flow meter.

Improper grounding is a common source of erratic magnetic flow meter readings. Even where adjacent pipe is conductive and serves as a ground, bonding jumpers can corrode.

Although used in the past, the trend is now not to furnish ultrasonic power generators to clean electrodes (Cooper, 1992). Ultrasonic cleaning is more appropriate for crystalline substances than for greases and biological growths found in wastewater pumping stations. However, the designer can select a magnetic flow meter, which can be retrofitted with either a permanent or portable ultrasonic power generator.

Compared to ultrasonic meters, magnetic flow meters are more expensive in larger line sizes. They are inserted directly to and actually substitute for the pipeline, compared to many ultrasonic models that clamp to the outside of the pipe.

Frequently, magnetic flow meter size is smaller than the connecting process pipe because the meter is designed with a minimum velocity to keep debris suspended and reduce electrode coating. As a guideline, this minimum velocity typically might be 0.6 to 0.9 m/s (2 to 3 ft/sec) for stormwater and raw wastewater applications (deGlanville, 1992). Under this guideline, for 10:1 flow ranges, maximum flows produce line velocities of 6 to 9 m/s (20 to 30 ft/sec), which may be hydraulically unacceptable. An alternate guideline is that minimum velocities of 0.15 to 0.3 m/s (0.5 to 1.0 ft/sec) are acceptable, provided flushing velocities of at least 1.8 m/s (6 ft/sec) are present daily (Gunby, 1993).

Providing bypass piping around a magnetic flow meter should be considered, especially in a pumping station where downtime cannot be tolerated. If a bypass is used, an access flange to allow cleaning the inside of the meter should be evaluated. In noncritical stations, the magnetic flow meter taken out of service might be replaced with a spool piece.

Field replaceable electrodes are not widely used in magnetic flow meters measuring stormwater and wastewater flow but may be beneficial in some applications. Electrodes have been known to develop coats and lose accuracy in some uses (e.g., with thick sludges). For stormwater and wastewater applications, comparable coating may take years to occur, if ever. However, certain industrial wastewaters can foul the electrode quickly. Thus, especially for large meters with routinely low flowrates, field replaceable electrodes may be worth the extra expense in certain applications. For the same reason, "electrodeless" magnetic flow meters might also be beneficial (Gemin, 1993).

The designer should be aware of the term *field replaceable*, which is used somewhat ambiguously. Most operations personnel desire removing the electrode from outside the magnetic flow meter, preferably under process conditions. If so, a field replaceable electrode with a hot tap assembly is needed. By contrast, some field replaceable electrodes require that the electrode be removed from inside the meter, which involves shutting down the process and expending labor.

Figure 7.4 depicts a magnetic flow meter schematic. Figure 7.5 shows a horizontal installation as recommended by one manufacturer. Note that a

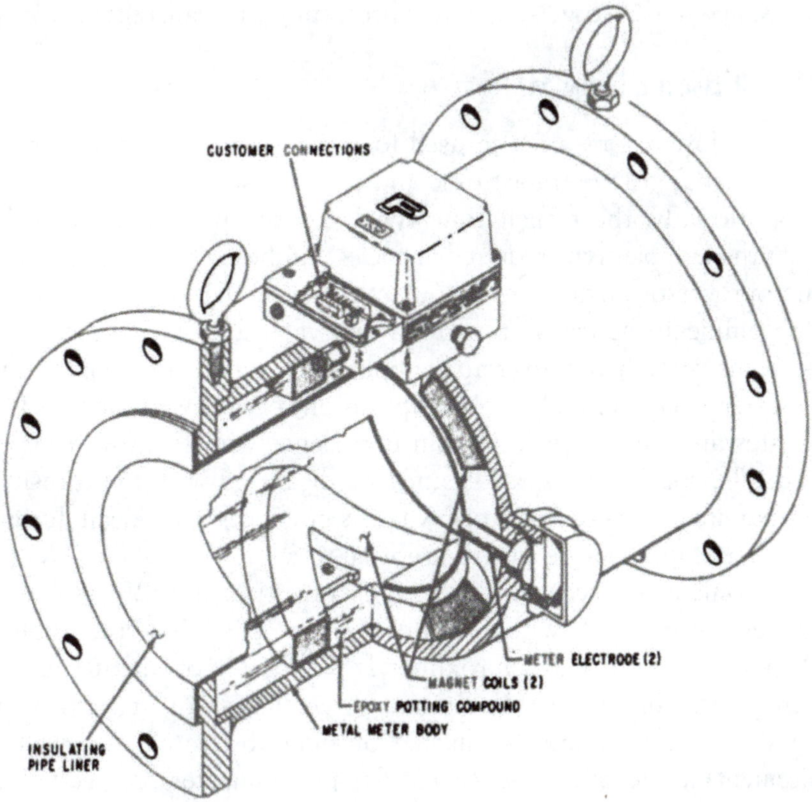

FIGURE 7.4 Magnetic Flow Meter Schematic

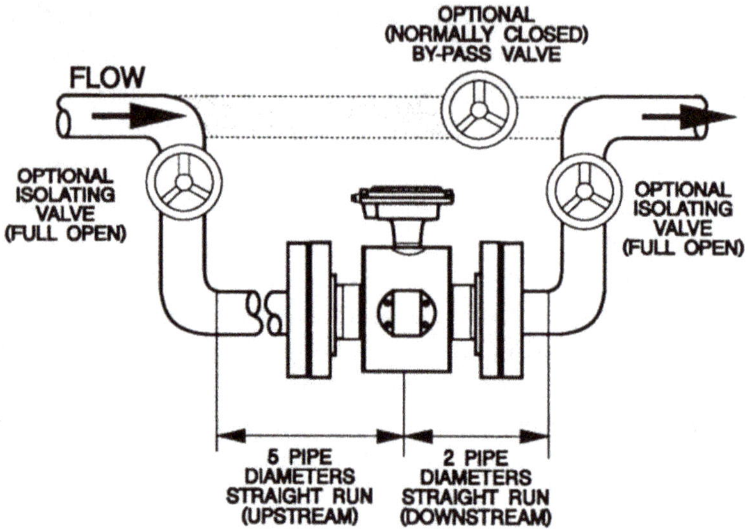

FIGURE 7.5 Magnetic Flow Meter Horizontal Installation

full magnetic flow meter is ensured. Figure 7.6 shows a vertical installation. Note the optional access flange for cleaning. Field-installed meters may have a solar shield and a surge suppressor protecting the transmitter.

3.2 Ultrasonic Flow Meters

Ultrasonic flow meters are also used for wastewater applications, and the two basic types are the transit time and the Doppler.

Traditionally, the transit time type required primarily clean fluids, whereas the Doppler type required particles, bubbles, or turbulence. Because of microprocessor advances, some manufacturers that supply both meters now recommend the transit time for wastewater and stormwater applications. In any case, there is overlap in applications where both might be used. Wastewater is a common Doppler application. The transit time type is used for wastewater and stormwater, but experience with stormwater is more limited. The specific application must be discussed with the transit time meter manufacturer because stormwater is not necessarily clean during the first flush and may not be clean for a long time.

Ultrasonic meter accuracy typically is advertised as ±2% of full scale. Some practitioners agree with this value (Gemin, 1993), but others (Guagno, 1992) put the accuracy at approximately 5% to 8% of rate. Range is not explicitly stated in much of the manufacturers' literature, in part because it changes as the meter's span is adjusted. Straight upstream and downstream requirements are ideally 10 to 20 and five pipe diameters, respectively. But some practitioners have had satisfactory performance by following the same

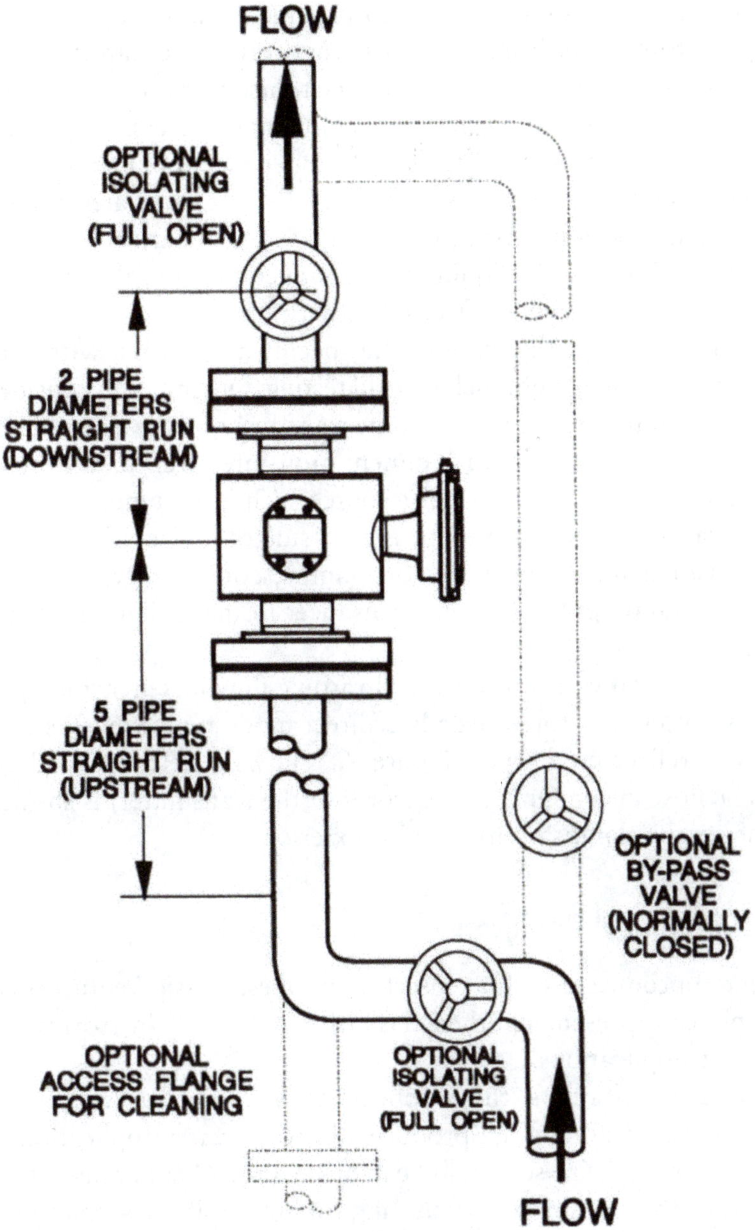

FIGURE 7.6 Magnetic Flow Meter Vertical Installation

laying lengths as recommended for the classic Venturi (Gemin, 1993). Minimum measurable velocity is typically 0.15 m/s (0.5 ft/sec).

A main advantage of ultrasonic meters is that in many models the transducer either clamps on or is track mounted to the pipe exterior and thus is nonintrusive. Pipe material must be able to conduct sound. Lined pipes may work, depending on the meter manufacturer and the extent of bonding.

The conservative design choice is to provide an unlined spool piece for subsequent transducer installation. Alternatively, some transit time designs include a line-size spool piece in which the transducer is integrally mounted.

From an installed cost standpoint, ultrasonic meters are attractive for larger line sizes because transducer size is independent of line size. Although ultrasonic meters have been used successfully in wastewater applications, success has not been uniform (Robinson, 1986). In part, this is because of variations in flow profile, liquid sonic conductivity, and the amount of air bubbles or particles, which affect reflectivity.

The user should fully explore the specific application with the meter manufacturer, which might include field testing. Other design considerations are to reduce pipe vibration, mount the transducer on the side of horizontal pipes to reduce foaming and sedimentation interference, and mount the transducer head away from a noise source such as a pump.

The transducer is mounted to the outside of a clean, bare metal pipe by means of a mounting track or clamps, coupling grease and metal straps. A common problem is the transducer losing its bonding to the pipe (Ouzts, 1992).

Figure 7.7 shows a transit time ultrasonic flow meter, including sensors and transmitter. The top sensor is a direct mode type, whereas the lower sensor is a reflect mode type. Figure 7.8 shows a field installed Doppler ultrasonic flow meter. Only the sensor (not the transmitter) is shown. Note how the sensor clamps on to the pipe exterior.

3.3 Differential Producers

Differential producers such as insert flow tubes, classic Venturi tubes, and orifice plates represent another class of meters used in stormwater and wastewater applications.

Among practitioners, there appears to be a wide difference of opinion on the use of differential producers in wastewater applications. In the opinion of some, their use should be avoided altogether because of clogging problems of the transmitter connecting tubing, capillary system problems, and the concomitant increase in maintenance. Other practitioners have had successful experience and unequivocally recommend them.

Accordingly, the practitioner considering differential producers should carefully evaluate them. Manufacturers should be contacted directly for their recommendations. Information on existing wastewater installations should be gathered to determine which hardware and installation practices are successful.

The most common differential producers used in wastewater and stormwater are the classic Venturi tube and the flow tube, an insert version of

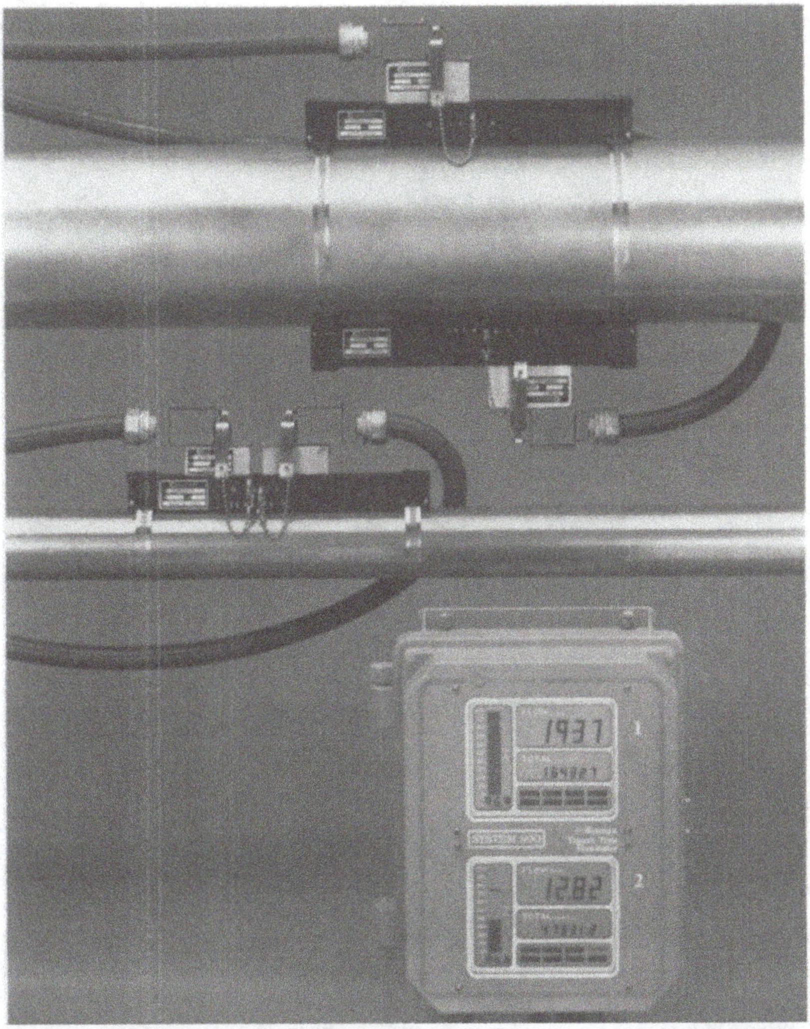

FIGURE 7.7 Transit Time Ultrasonic Flow Meter

the classic Venturi. The least expensive differential producer is the orifice plate, but its use is limited in wastewater and stormwater applications. It is energy inefficient because most of its differential pressure is nonrecoverable and it is more susceptible to debris collection.

Some classic Venturi meters can measure bidirectional flow, which may be of use in some applications. Because it is less costly, the flow tube is more often used than the classic Venturi. Although the classic Venturi typically has a circular cross-sectional area, rectangular cross-section Venturi meters are available to mate with rectangular concrete pipe sometimes found in larger pumping stations.

FIGURE 7.8 Doppler Ultrasonic Flow Meter

Differential producers share the disadvantage that their pressure taps and connecting tubing between the meter itself and the electronic transmitter tend to become clogged from debris and solids. To combat this, water purges and sealed capillary systems are used.

Water purges can be either continuous or intermittent. Continuous purges decrease system accuracy by approximately 1% of rate because the purge water adds backpressure to the differential pressure measurement (Briggs, 1993). Intermittent purges have been used successfully (Ouzts,

1992). Intermittent purges can be initiated either manually or automatically at periodic time intervals. However, intermittent purges produce intermittent "blips" in the flow data. This typically is not a concern in wastewater and stormwater.

In sealed capillary systems, the connecting tubing is fluid filled and isolated from the process fluid by the seal. In the past, users of sealed capillary systems have experienced problems of inaccuracy because of differential temperature heating of the two capillaries, increased inaccuracy at low flows, and calibration difficulties. To alleviate these problems, capillary tubes should be bundled together to reduce temperature gradients. Some manufacturers are now offering modified seal designs. More convenient calibration techniques are also becoming available.

Overall system accuracy for different producers is in the range of ±0.25% to 5.0% of rate depending on meter type, flow turndown, bore calculation errors, errors because of the capillary system (if used), and errors of the differential pressure transmitter.

Because flow varies as the square of differential pressure, the accuracy of differential pressure measurement limits flow turndown to only approximately 3.5 : 1 (newer, more accurate designs such as "smart" transmitters increase the turndown). Such low turndown promotes designs in which a flow meter is dedicated to each pump because individual pump turndown is less than the overall turndown of the pumping station.

To increase turndown, sometimes two differential pressure transmitters with overlapping ranges are connected to one meter. An attendant electronics unit switches to the transmitter of the appropriate range. A proprietary unit that accomplishes this switching is available.

The coefficient of discharge for a differential producer changes below a certain Reynolds number (approximately 70 000 to 80 000), and inaccuracy will result. Accordingly, the Reynolds number should be checked, particularly at low flows.

Most of the differential produced by the classic Venturi or flow tube is recoverable, unlike an orifice plate, which has much greater nonrecoverable losses. Typically, the non-recovered head in the classic Venturi or flow tube is approximately 5%. For water applications, a typical, recoverable head loss at maximum flow for the classic Venturi or flow tube is approximately 25 kPa (100 in. water [H_2O]). Thus, the non-recovered or lost differential is only approximately 1.2 kPa (5 in. H_2O).

Typically, the head permanently lost by the classic Venturi or flow tube by itself is inconsequential. However, these tubes sometimes need to be downsized to develop the required differential pressure for accurate measurement. Although more of a concern in gravity systems, the fittings required for the downsizing create a loss that can be substantial to the point of affecting pump selection. Accordingly, Venturi or flow tube selection should occur early in the design so it can be incorporated to pump selection.

Throats of differential producers used in stormwater and wastewater applications can be either bronze or stainless steel. Stainless steel is the conservative design choice. Throat dimensions are integral to producing accurate data. Over time, stainless steel throats might maintain their dimensions more than bronze, which is softer and may be more prone to galvanic corrosion because it is more dissimilar to the ferric tube or pipe. Stainless steel is attacked by chlorine and, if chlorine is expected, the application should be reviewed with the manufacturer.

Regardless of the differential producer used, the differential-measuring transmitter should be located below the pressure taps and the connecting tubing should be gently sloped. Entrapped air bubbles will thus be reduced, although lines should still be bled periodically. In cold weather areas, connecting tubing located outdoors should be both heat traced and insulated.

The required straight run of upstream pipe can be substantial and depends on the type of differential meter, the beta ratio (diameter of throat to inside pipe diameter), and the type of flow disturbance (e.g., elbow or valve). A classic Venturi tube of β 0.6 requires seven straight upstream diameters if the upstream obstruction is a reducer. If the obstruction is two elbows not in the same plane, the required straight upstream length increases to 16 diameters.

For isolation valves upstream of the differential producers, some manufacturers specifically recommend against the use of butterfly valves (DiNapoli, 1993). In its fully open position, the butterfly valve is in the middle of the flow stream and introduces hydraulic disturbances that might affect the meter's performance.

Major manufacturers of differential producers have their own tables and curves, which recommend required lengths of straight upstream and downstream diameters for various hydraulic disturbances. These recommendations are based on actual test data. Some differential producers are fabricated by minor manufacturers that do not perform similar tests. Obtaining meters without supporting test data increases the risk of poor flow measurement.

A proprietary differential producer exists whose typical applications have been for wastewater sludges, but this device may not be applicable for wastewater or stormwater flow. A spool of concentric circular cross section is cast with a wedge-shaped "dent" on the top half that induces a pressure differential across the wedge. Wafers with pressure taps are cast into the spool on either side of the wedge. A closed capillary system transfers the differential pressure to a transmitter. Flow path is straight through the bottom of the spool, so no solids settle out or accumulate. Pressure recovery downstream is good.

Figure 7.9 shows a continuous purge assembly for a Venturi meter.

3.4 Open Channel Flow Meters—Primary Devices

Primary devices commonly used to measure open channel flow include Parshall flumes, Palmer–Bowlus flumes, and open flow nozzles. Although these devices all exhibit a loss of head, this is typically less of a concern in pumping station design than in pure gravity systems.

Parshall flumes consist of a converging section, a throat section, a diverging section, and a stilling or measurement well. There is a decrease in level as water passes through the flume. Flow is nominally proportional to the three-halves power of the level change. *Instrument Engineers' Handbook—Process Measurement* (Liptak & Venczel, 1982) states that this device is not extremely accurate but is adequate for wastewater applications. A commonly assumed accuracy is ±5% of rate (Briggs, 1993).

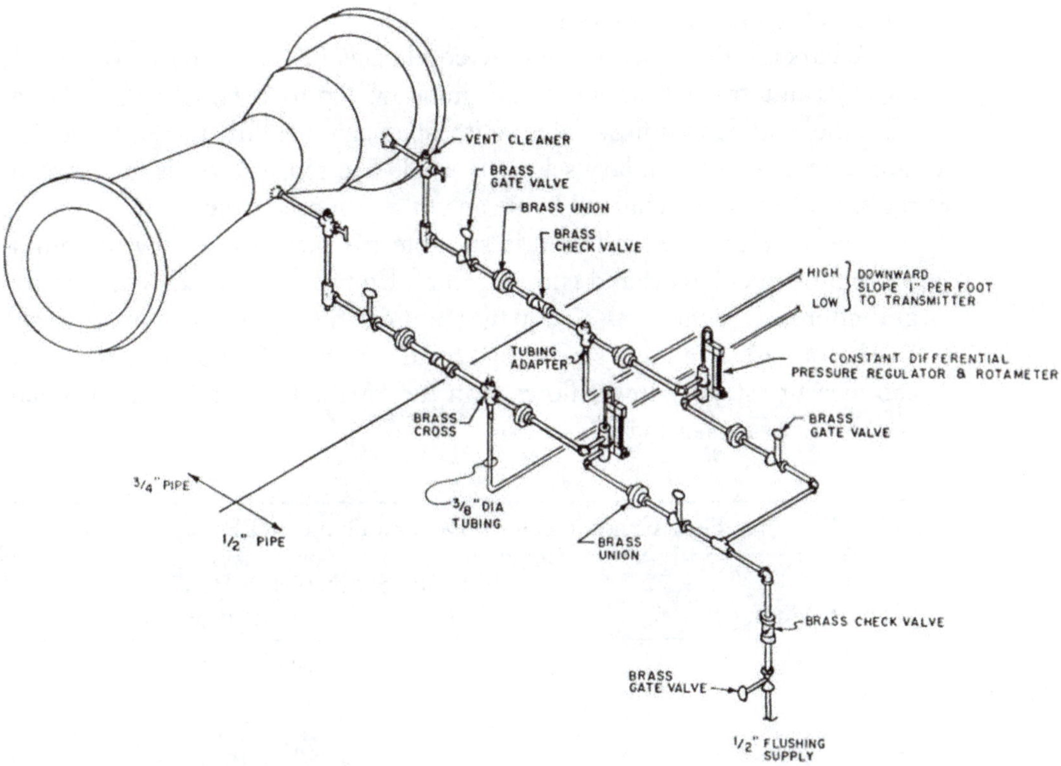

FIGURE 7.9 Continuous Purge Assembly for a Venturi Meter (ft × 0.3048 = m and in. × 25.4 = mm)

Parshall flumes are dependable, require minimum maintenance, and tend to be self-cleaning because high velocities flush out solids. Available throat widths range from 25 mm to 3 m (1 in. to 10 ft). Flow range for a given flume can be as high as 35:1, but, in practice, the secondary's range limits the flow range to much less.

Typically, Parshall flumes are designed for free-flow (no backwater) conditions in the diverging section. As flow increases, backwater develops in the diverging section, which eventually affects accurate measurement. In some instances, such as retrofits constrained by existing hydraulics, backwater conditions may be unavoidable at certain flows. If such flows are of interest, level measurement in the diverging and the converging section must be measured. This two-point measurement adds complexity to calculating flow; a few small vendors specialize in this type of measurement.

The Parshall flume should not be installed near turbulent, surging, or unbalanced flow, or near a poorly distributed velocity profile. It should be located without bends immediately upstream.

A drawback of Parshall flumes is their expense, primarily in the concrete work needed for the flume itself and the additional concrete work needed upstream to ensure a smooth transition.

A Parshall flume with free-flow conditions (as indicated in Table 7.1) includes an ultrasonic sensor measuring flow depths at H_a and H_b and converts the reading to a flow value. With open channel flow measurement, an admittance-type secondary sensor is used. The sensor strip is mounted to the side of the open channel by means of a mounting track.

The Palmer–Bowlus flume is suitable for measuring flow in round-bottom open channels and is trapezoidal. Units are available with an integral approach section. Its advantages and disadvantages are similar to the Parshall flume. Available sizes range from 100 to 750 mm (5 to 30 in.). It can measure slightly lower flows than the Parshall flume, but the Parshall is capable of measuring higher flows.

TABLE 7.1 Free-Flow Conditions for a Parshall Flume (WEF et al., 1992)

Throat Width, cm	Maximum Submergence $H_a/H_b \times 100$ for Free-Flow Condition, %
≤ 7.6	50
15.2	60
22.9	60
$> 22.9 < 30.5$	70
≤ 30.5	67

Note. cm × 2.54 = in.

Weirs are the most accurate of these devices but are seldom used in wastewater or stormwater systems because of their tendency to allow solids and debris to settle out at the weir base.

Flow nozzles include the Kennison nozzle and at least one proprietary device. These devices measure flow in open channels or partially filled pipes. The devices are metal with a flanged circular inlet and there is free discharge. The proprietary device has an optional transition section, which allows connection to a downstream pipe.

Nozzles are available from 150 to 910 mm (5 to 35 in.) and with associated flow spans from 125 to 875 L/s (200 gpm to 20 mgd). Stated accuracies are at least as good as for Parshall flumes.

These devices have an intermediate section, the geometry of which causes a change in water surface level, which is the basis of flow measurement. Secondaries for these devices include ultrasonics, admittance probes, float systems, and even pressure-sensing systems. Required straight upstream diameters and other installation requirements should be reviewed with the manufacturer.

These devices are similar to Parshall flumes, including their self-cleaning feature. For some installations, a potential advantage is a lower total cost over a Parshall flume because of a savings in concrete formwork.

3.5 Open Channel Flow Meters—Secondary Devices

Common types of electronics that measure level and convert it to flow are ultrasonics and admittance devices. Float-actuated linkages are now less common, probably because of their reliance on moving parts and their requirement of a stilling well, which tends to collect solids. The ultrasonic transducer is virtually identical to that used for ultrasonic level measurement. The admittance device transmits radio frequency to measure capacitance as a function of level. The device has circuitry, which reduces the effect of electrical paths created by process buildup on the probe. Proper grounding is necessary, and a grounding plate is typically provided.

3.6 Open Channel Flow Meters—Accuracy

The accuracy of open channel flow meters is open to some speculation. For weirs, with proper precautions, the relation between flow and level is accurate to within 2% (Liptak & Venczel, 1982). Parshall and Palmer–Bowlus flumes are less accurate. Although an accuracy of approximately 5% is assumed, field tests and unpublished studies indicate greater inaccuracy in at least some installations (Briggs, 1993).

Manufacturer information suggests secondary ultrasonic devices measure level with accuracy greater than 1%. Equivalent information for secondary admittance devices is more difficult to find but is comparable. Field observations and common sense suggest open channel flow measurement is often more inaccurate than commonly assumed.

At most, head height is only 0.4 m (1.25 ft) and is often much less. An error in liquid measurement of, for example, 13 mm (0.5 in.) affects accuracy substantially. Potential sources for error are numerous. Foam (even slight) in wastewater and debris collected along walls and edges can change the surface profile. Wind can produce ripples. Other error sources include excessive approach velocities, deviations in installing the flume liner, and inlet flow, the profile of which has still not transitioned from pipe to open channel flow.

Flow inaccuracy is the sum of the inaccuracies of the devices that comprise the flow-measuring system. For open channel flow meters and differential producers, one must account for the inaccuracies of the primary and secondary elements. The secondary might be miscalibrated and can drift. Admittance devices appear prone to some drift in wastewater applications, possibly from coatings buildup. Periodic maintenance is necessary.

Pumping stations do have the benefit of a wet well. If removing the pumping station from service is permitted, the wet well can be used to verify meter accuracy through draw-up calibration tests.

3.7 Ancillary Flow Meters

Other flow meters are used to support ancillary functions of the wastewater or stormwater pumping station. Rotameters and devices that visually indicate only flow presence are used on small flows such as for seal water. Rotameters are also used to measure airflow in bubbler systems. Propeller meters are often used for applications that require larger amounts of clean water, such as washdown water.

4.0 PRESSURE MEASUREMENT

Compared to water distribution pumping stations, pressure measurement is of less concern in wastewater and stormwater pumping stations.

Pressure gauges are typically furnished on the discharge and suction of each pump. An isolation valve typically is installed with the gauge or seal to allow easy removal. Placing gauges on both pump suction and discharge assist in determining the actual field pump curve. Any pressure gauge on the pump suction should be provided with a combination vacuum or pressure range. Pressure switches are less prevalent and are used for such applications as monitoring seal water pressure and pump discharge pressure (to confirm the pump is operating).

Pressure gauges typically are selected so that indication is between 30% and 70% during normal operation. Ideally, the pressure gauge should indicate less than 100% of full scale at the shutoff head. Often, both conditions are not possible, and a compromise must be reached. Most pressure gauges can be over-pressured approximately 30% without needing to recalibrate the gauge.

Occasionally, only one gauge per pumping station is used. In such cases, process connections (either plugged or furnished with an isolation valve) should be installed on both pump suction and discharge to facilitate adding pressure gauges if they are found to be needed.

Pressure gauges tend to fail prematurely from mechanical shock from hydraulic transients and fast-closing valves. Accordingly, gauges are furnished with pulsation dampeners, are liquid filled, or are remotely mounted from the process pipe.

Pressure gauges typically are fitted with quarter-turn isolation valves. To extend gauge life, the valves should be closed unless a reading is being taken.

For protection from debris and the process fluid itself, gauges are often furnished with diaphragm seals filled with an inert fluid. Typically, the seals themselves are sufficient, but as a supplement, water purges on the process side of the seal can also be used. A suitable diaphragm material is an appropriate stainless steel alloy. A 13-mm (0.5-in.) National Pipe Thread process connection on the seal generally is satisfactory. If excessive debris or grease is anticipated, 20-mm or even 25-mm (0.75-in. or even 1-in.) process connections can be used.

Two devices (gauge plus switch) mounted on one seal is common practice. Mounting three (or more) devices on one seal should be reviewed with the manufacturer. The diaphragm-seal fill fluid may be unable to provide enough displacement volume. A slow-responding or inaccurate device may result, and the physical mounting of three devices may prove troublesome.

An alternative to diaphragm seals are "donut-shaped" annular seals mounted off the main process line. A typical size (of the "donut's hole") would be 25-mm (1-in.) Female National Pipe Thread, and these seals can be flushed clean by opening an isolation valve. Mounting annular seals offline is still not common and thus experience is limited. Traditionally, annular seals have been mounted inline in the main process line. Because wastewater and stormwater are relatively low in solids, mounting annular seals inline probably is overdesign. For larger pipe diameters, cost becomes a factor because annular seal size is the same as the process pipe.

Although not routinely used in wastewater and stormwater pumping stations, pressure switches occasionally have some application. Normally, process pressure conditions are relatively low (less than 345 kPa [50 psig]). Unfortunately, many standard commercially available switches in this range have appreciable deadbands, which can lead to nonreproducible behavior and control strategy malfunction (Gemin, 1993). In these cases, either a nonstandard expensive pressure switch must be used or the control strategy may need to be abandoned altogether.

Figure 7.10 shows an installed pressure gauge and diaphragm seal assembly, with the diaphragm seal directly below the gauge.

5.0 ALARMS AND INDICATION

5.1 Alarm

In small pumping stations, only abnormal high and low levels might be alarmed. Many existing small pumping stations do not alarm abnormal low level. However, this practice is not recommended because it provides no alert that a pump is running dry.

FIGURE 7.10 Pressure Gauge and Diaphragm Seal Assembly

For larger sophisticated pumping stations, alarms can become numerous, and their use should be evaluated carefully. Many designers provide for monitoring multiple alarms locally at the pumping station and remotely at a central facility. In some applications, or in instances where a cellular or auto dialer system is used, the designer may choose to only send a single station trouble alarm (aka common alarm) to a central facility.

Table 7.2 shows the numerous conditions that might be alarmed at larger pumping stations. Personnel safety pushbutton alarm stations have been installed no more than 50 mm (24 in.) above the floor to allow an injured operator to call for help.

Pump failure is primarily detected by high liquid level in the wet well. In more sophisticated pumping systems, secondary methods to detect pump failure also are used.

There are several common secondary methods of detecting pump failure. In each, the control system monitors that the pump has been proved to be

TABLE 7.2 Selected Alarm Conditions at Wastewater and Stormwater Pumping Stations

Condition	Condition
• (Each) pump fail to run[a]	• Low ventilation airflow
• (Each) pump low seal water pressure	• Building security
• (Each) check valve fail to open	• Building drain sump high level
• High wet well level	• High motor temperature
• Low wet well level	• Power failure
• High discharge flow	• Emergency generator in operation
• Low discharge flow	• Intrusion alarms on perimeter fencing
• (Each) pump high vibration	• (Each) check valve fails to close
• (Each) pump high bearing temperature	• High building temperature
• Moisture present (within submersible pump motor)	• Low fuel
• Combustible gas present	• Phase failure
• Oxygen deficiency	• Motor/breaker failure
• Variable frequency drive diagnostic	• Surge fail to close
• Remote terminal unit failure	• Building flooded
• Surge valve open	

[a]See text for ways of sensing pump fail

on within a preset (but adjustable) time. If no proof is received, a pump failure is assumed. Although not definitive proof, a trouble-free means of sensing pump operation are the "M" contacts from the pump motor starter. More definitive but still a secondary means of sensing a running pump include check valve limit switches, flow monitoring of the pump discharge, and pump discharge pressure. Check valve limit switches are reliable if the check valve is so designed. Field installing limit switches on check valves not designed to be furnished with such is problematic. Suitable, reasonable-cost flow switches for "dirty water" applications are hard to find. If used, pressure switches should be located on the discharge between the pump and check valve. Sometimes, the correct pressure setpoint is difficult to determine. Contributing factors include wide changes in suction pressure, low dynamic friction pipe losses, and multiple-pump operation that changes normal operating pressure.

Often, hand switches that control pumps are three-position switches (ON/OFF/AUTO or LOCAL/OFF/REMOTE). These fail conditions typically are interlocked regardless of the position of the selector switch. To avoid nuisance tripping during testing or maintenance, some designers will include a bypass selector switch around the alarm condition that can disable shutting the pump off if needed. As an alternative, some designers may choose to have the fail conditions interlocked only with the selector switch in the AUTO position, so that nuisance alarms might occur if the switch is operated in the ON position during testing or maintenance. Some designers choose to only have the alarm provide an alarm and not stop the pump, leaving the decision to the operator who may be monitoring the equipment remotely if the equipment should be shut down.

An exception to this rule is if the fail condition can cause injury or excessive mechanical damage. For these critical fail conditions, some designers choose to provide an alarm and an interlock to stop the pump even in the ON mode. An example is stopping a progressive cavity pump on high pressure.

Atmospheric monitoring devices typically are installed in pumping station working areas. The requirements of National Fire Protection Association (NFPA) 820 (NFPA, 2020) should be followed, but the party that typically needs to be satisfied is the local fire marshal, whose requirements may differ. National Fire Protection Association 820 requires that combustible gas detectors be placed in rooms or enclosed spaces of both wastewater and stormwater pumping stations, and that they be of the failsafe type. Also, NFPA 820 requires that loss of ventilation flow be monitored. In general, NFPA 820 requires that these alarms be both visual and audible. Other devices often installed include combustible gas and oxygen-deficiency detectors.

5.2 Equipment Status Indication

The status of most major process equipment is indicated at the pumping station. Typical status indications include those for pumps (constant speed), variable-speed drives, check valves, screening, and de-gritting equipment (if present), ventilation fans, station power, and emergency generator. Occasionally, the pumps commanded to run in the automatic sequence (lead or lag) are also indicated.

With the exception of smaller pumps, elapsed run time of each pump since installation is typically indicated at the pumping station or at the central facility. The run time and number of starts since midnight (or some other time) might also be indicated.

Above a certain motor size, for example, 37 kW (50 hp), motor amperage draw, voltage to each of the three legs, and motor speed (in revolutions per minute) typically are indicated, particularly with variable-speed drives.

For larger sophisticated pump stations, the use of a data highway connection between devices allows for multiple variables to be monitored and controlled with the use of a single cable connection. Some designers will use the data highway connection to monitor basic equipment status. Monitoring data for a variable-frequency drive (VFD) could include run status, drive running hours, drive line harmonic voltage and power factor, motor average voltage, motor average ampacity, speed feedback, and trip. Other data such as the cause of the last trip and the last date and time of the trip can also be pulled from a data highway connection.

Data highway connections can also be provided to other equipment such as generators, transfer switches, power meters, protective relays, solid-state soft starters, and smart or intelligent motor control centers. Some instrumentation also can communicate via the data highway. Each data highway connection can pull hundreds of data points; therefore, consideration should be given that an operator is not overloaded with too much information that it becomes overwhelming. There are numerous communication protocols that are available, and it is possible a protocol converter may be needed if a device uses a different communication protocol than what is used by the pump station's master controller.

5.3 Process Variable Indication

Depending on the pumping station and overall system complexity, the number of variables measured can be none to several. Some guidelines apply. Flow, wet well level, and totalized flow typically are indicated at the pumping station. Flow is more often recorded than level.

Some existing pumping stations use recorders. The use of recorders at the pumping station is sometimes problematic. Recorders need pen and chart paper replacement and are prone to attack by corrosive gas associated with wastewater pumping stations. Accordingly, recorders should be installed in a protected atmosphere to be reliable and long-lasting. If located outside, recorders are subject to environmental attack, such as by sunlight and heat. To reduce environmental effects, recorders should be furnished with a rain and sun shield and should be installed facing north.

More common is for pumping stations to include local transmitters that include a digital display. The transmitter can be located near the field sensor but can also be remote mounted at a more accessible location. For example, a level transmitter can be located on top of the tank where the sensor is located, or it can be mounted just outside the tank at ground level on a rack or located in an electrical room. If located outside and exposed to sunlight and heat, then a rain and sun shield should be installed with the transmitter facing north. If remote mounting the transmitter, the designer should verify the cable distance limitations with the manufacturer.

When monitoring flow, the analog output signal from the flow meter can be used to monitor real-time flow as well as calculating totalized flow. To minimize a compounding summation error due to the signal losses associated with analog signals, some flow meters include the ability for two output signals: an analog output and a discrete pulse signal that will pulse after so many gallons to totalize flow. If available, the designer can use a data highway connection instead of an analog and discrete pulse signal.

For supervisory control and data acquisition (SCADA) systems, the central facility typically has a report printer and desktop monitor. The report printer sometimes is omitted, which can be a problem if a hard copy of reports is desired.

6.0 OPERATOR INTERFACES

Operator interfaces at the pumping station site include a control panel and hand switches near the pump. The pumping station could also include a graphic user interface with a keypad and/or touchscreen at the control panel and/or local SCADA cabinet. At remote central facilities for SCADA telemetry systems, the operator interface typically is a desktop workstation with keyboard.

The face of the control panel at the pumping station typically includes hand switches, status-indicating lights, alarm lights or an annunciator, and process variable indicators and recorders. Operator interface units typically are provided on the panel face for devices such as single-loop digital controllers (SLDCs) and programmable controllers. Some existing simple stations that use programmable controllers may not include an operator interface. Several manufacturer pump controllers now include a graphic user interface that will be mounted on the face of the control panel, where an operator can monitor equipment status and modify the programmable setpoints.

In the last decade, the use of local graphic user interfaces is becoming more common. These displays can vary in size (commonly 0.2 to 0.5 m [7 to 19 in.]). Some designers will include a graphic user interface at the local control or SCADA cabinet for even simple pumping stations and not just for more sophisticated stations. The graphic user interface at the local SCADA cabinet can be used for monitoring only or to both monitor and control the station. It can also display information from other remote sites if connected to the overall SCADA system. If a local graphic screen is provided, the designer needs to consider the location and temperature limitations of the device. If located outside and exposed to the sun and other elements, a hinge cover is recommended or the designer can consider placing the graphic screen inside the inner door of the cabinet.

A local personal computer occasionally is used to access the programmable controller to examine registers, modify strategies, and adjust parameters. Some equipment (e.g., VFDs and local SCADA programmable logic controllers [PLCs]) can be accessed by using a personal laptop. Some designers will include a foldout laptop tray at the cabinet for the operator's use.

6.1 Control Panels

Control panels should be constructed with the appropriate National Electrical Manufacturers Association (NEMA) rating. Table 7.3 describes NEMA classifications for control panels typically used in wastewater and stormwater pumping stations. The listed panels are rated for use in nonhazardous locations; panels rated for hazardous use are seldom required in wastewater and stormwater pumping stations.

Because NEMA 12 panels do not resist corrosion well, the more conservative and appropriate choice for indoor process areas in wastewater pumping stations is NEMA 4X (watertight and corrosion resistant). Along with NEMA 1, NEMA 12 might be appropriate in control and in electrical rooms. Where high corrosion is expected or has been experienced, NEMA 4X panels have been used. For outdoor applications, the conservative choice is NEMA 4X panels. In areas that are not humid and with lesser amounts of corrosive sewer gases, NEMA 3, NEMA 3R, and NEMA 3S have proved adequate.

The corrosion test that a NEMA 4X panel must pass involves subjecting the panel to a salt spray. Thus, NEMA 4X panels should not be considered

TABLE 7.3 National Electrical Manufacturers Association Classifications of Control Panels Commonly Used in Wastewater and Stormwater Pumping Stations (Adapted from American National Standards Institute/NEMA 250, Enclosures for Electrical Equipment, 2020)

Enclosure	Description
NEMA 1	Indoor use—some protection against contact with the enclosed equipment
NEMA 3	Outdoor use—protection against windblown dust, rain, sleet, and external ice
NEMA 3R	Outdoor use—protection against falling rain, sleet, and external ice
NEMA 3S	Outdoor use—protection against windblown dust, rain, sleet; operation of external mechanisms when ice laden
NEMA 4	Indoor or outdoor use—protection against windblown dust and rain, splashing water, and hose-directed water
NEMA 4X	Indoor or outdoor use—protection against corrosion, windblown dust and rain, splashing water, and hose-directed water
NEMA 12	Indoor use—protection against dust, falling dirt, and dropping noncorrosive liquids

universally corrosion resistant. Common enclosure materials for NEMA 4X panels include aluminum, fiberglass, 304 stainless steel and 316 stainless steel.

The NEMA 4X panels add a complication in that many analog operator interface devices such as indicators and recorders are not similarly rated. Accordingly, Plexiglas™ windows must be provided on the panel front for these devices. Corrosion-inhibiting capsules should be placed within the control panel, though a dehumidification system also may prevent corrosion.

For outdoor installations, solar shields should be used on control panels in high solar energy environments, and within which are housed sensitive solid-state electronics. Instead of solar shields, some practitioners have painted the control panel exterior white to reflect solar radiation. Such control panels should be placed with their face in the northerly direction to reduce heat buildup and to allow easy viewing of low-intensity light-emitting device indicators. Even in relatively warm areas, control panels located outdoors should have thermostatically controlled heaters to prevent condensation.

Control power is typically 115 V–ac. In some applications, low-voltage (24-V–ac or –dc) control power is used instead. Examples of these applications include standby engines or certain VFDs, for which low-voltage control power is the manufacturer's standard. Such low-voltage control circuits should be used with caution. Electronic devices in low-voltage control circuits appear more prone to induced voltages from nearby process equipment power—460-V, three-phase in the United States and 575-V, three-phase in Canada. The induced voltages sometimes render low-voltage control relays nonfunctional. "Band-aid" field fixes such as using interposing relays may work, but the interposing relays themselves require 120-V–ac power.

Figure 7.11 shows a stainless steel NEMA 4X control panel located outdoors.

6.2 Ancillary Instrumentation

Ancillary instrumentation includes devices such as elapsed time meters and process variable indicators, recorders, and totalizers. The enhanced capabilities of microprocessor-based circular chart recorders make them attractive. Multiple process variables can be recorded and otherwise processed. Chart

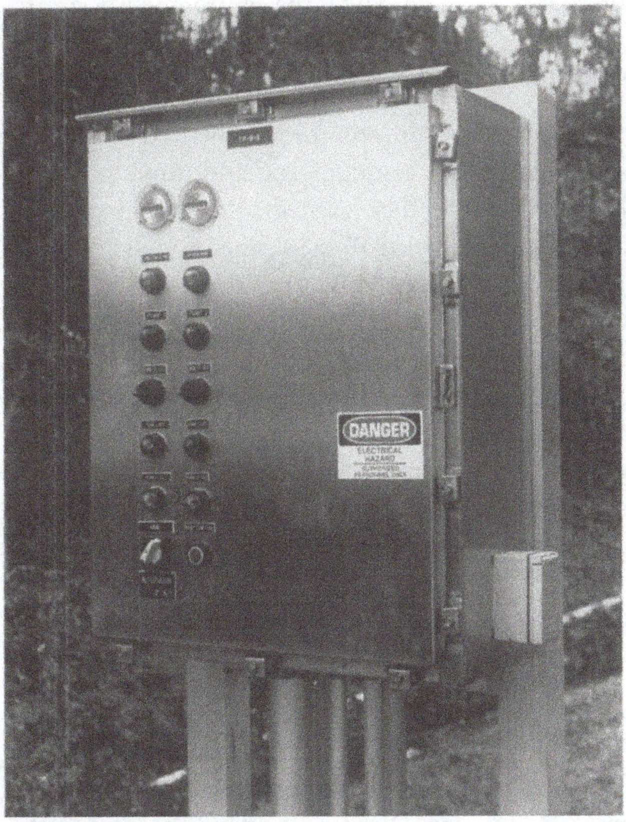

FIGURE 7.11 A NEMA 4X Control Panel

ranges are custom printed by the recorder itself. Also, process-variable totalizers are provided integral to the recorder.

7.0 CONTROL SYSTEMS

Types of control systems for automatic pump operation include conventional control relays, PLCs, SLDCs, proprietary controllers, and "smart" remote terminal units (RTUs).

Distributed control systems seldom are used in wastewater and stormwater pumping stations because they are suited for large, complicated processes. The application of personal computers typically is reserved for use at central facilities in SCADA.

The choice of control system depends on cost and control complexity and flexibility. The control systems typically are located at the pumping station and, if there is a central control facility, it generally performs "supervisory control" only with real-time control implemented by controls at the pumping station. Occasionally, central facility control is the normal mode, with reversion to local control only if the central facility control system fails.

7.1 Conventional Control Relays

Conventional control relays are still frequently used and are dependable and easy to obtain, even locally. Ladder logic and implementing logic changes with control relays usually can be learned without taking specialized courses.

Simple programmable controllers, acting only as relay replacers, sometimes have a lower installation cost than relays, though relays might still be appropriate in some cases. For instance, most practitioners would not want multiple pumping stations where each control system was composed of a different model, sometimes obscure, programmable controller lacking both proper documentation and programming units.

As control schemes become more complex, the amount of wiring and number of relays becomes cumbersome. Making field changes is time consuming, and programmable controllers become appropriate in these instances.

7.2 Programmable Logic Controllers

Programmable logic controllers have become increasingly prevalent because they are field programmable and implementing control logic is flexible. Properly installed PLCs have demonstrated high reliability.

Programmable controllers were originally developed as relay replacers in the automotive industry. Discrete sequencing logic is their forte, and programmable controllers are still typically programmed in ladder logic. Their ability to manipulate analog variables has increased to the point where they are trusted to perform relatively simple analog control functions. For more complex analog control functions, the amount of software programming and documentation becomes cumbersome.

For pumping station applications, programmable controllers typically start and stop pumps and control pump speed as a function of wet well level or to maintain a flow setpoint, alternate the pumps in sometimes complicated schemes, totalize flows, and monitor for fail conditions. Programmable controllers benefit from power conditioning, either with a power conditioner or an uninterruptible power supply (UPS), in which the designer wants the programmable controller to function during power failure. Most programmable controllers are rated for operating at temperatures between 0 and 60 °C (32 and 140 °F) and relative humidity between 5% and 95% without condensation. Thus, programmable controllers can be installed in areas other than air-conditioned control rooms if precautions (such as installing control panel heaters or solar shields) are taken. Programmable controllers have been known to halt their operation when exposed to too much sun or when located within a building during hot summer days.

Programmable controller output modules are classified into two basic types: a voltage output that can drive devices or an isolated contact output, which is a newer development. Proponents of either type can be found.

A suggested practice in using voltage output modules is to wire them directly to controlled devices, such as relays, which reside in the same control panel as the programmable controller. If the signal is to leave the panel, interposing relays should be used, which cuts down outside electrical interferences.

Isolated contact outputs can be directly wired into control circuits outside the panel and are available for driving up to NEMA 5 starters. The advantage of the isolated contact outputs is that they eliminate relays. However, the module itself is more costly. Also, because this module is larger, there is little if any savings in panel size. Besides a maximum current rating when driving a load, isolated contact outputs also have a minimum required current to extend contact life. This minimum current rating should be accounted for in the design.

7.3 Single-Loop Digital Controllers and Other Devices

Single-loop digital controllers are also field programmable and typically are suited for applications where less control logic is required. Though more suited for analog control functions, SLDCs also can provide discrete sequencing. Some SLDCs use a menu of logic blocks to implement the algorithm. Certain practitioners prefer this type of "language" over ladder logic.

Several proprietary control systems are also available. In one, a microprocessor-based controller uses preprogrammed software routines to control pumps for situations normally encountered in wastewater and stormwater pumping stations. Routines include ramping adjustable-speed pumps as a function of wet well level and various alternation schemes.

Supervisory control and data acquisition systems communicate with pumping stations by means of RTUs located at the pumping stations themselves. In years past, the only function of the RTU would be as part of the communication link. Local logic was done by other devices such as control relays or PLCs. "Smart" RTUs are increasing in popularity. Because local control algorithms reside in these devices, the design is simplified.

8.0 CONTROL ALGORITHMS

Control algorithms must achieve the process objective and also must not expose the pump to harmful operating regimes. Pumping stations can be composed of constant-speed pumps, variable-speed pumps, or a combination of the two.

The simplest process objective is to pump the wet well contents to a receiving point without overflowing the wet well. Constant-speed pumps and their algorithms are appropriate here. Also, some WRRFs prefer there always be some minimum flow to the facility. This is sometimes accomplished with a small, constant-speed "jockey" pump. As diurnal flows increase, the normal pumping scheme is invoked.

Often, the receiving point prefers accepting a varying flow not subject to step increases. An example is a large pumping station that discharges to a small WRRF.

Applicable algorithms require at least some variable-speed pumping capacity. Such variable-speed capacity is provided with all variable-speed pumps, a combination of variable- and constant-speed pumps, two-speed motors, or algorithms that make use of different-size pumps.

Control algorithms must also guard against potential harmful effects on pumps. For instance, as level rises and a large capacity, high head pump starts, a smaller capacity, low head pump might be interlocked to prevent it from pumping against its shutoff head and incurring damage.

There are numerous control schemes, equipment, and logic for pumping applications. A thorough analysis of the pumping station is required to confirm the most reliable and cost-effective system to use. The following subsections present some guidelines.

8.1 All Constant-Speed Pumps

In the most general case of an all-constant-speed pumping station, each pump is provided with unique start and stop level setpoints. A simple variation typically used on small pumping stations is to start multiple pumps individually on successive higher wet well levels, but to stop them at the same level. As pumping station size increases relative to the receiving WRRF, this variation places unnecessary flow and load surges on the facility and should be avoided.

Another variation is to use time-delay logic coupled with the same level setpoint to control multiple pumps. For instance, a lead pump will start on a rising high level, and the lag pump will start if high level is sustained for a preset time, such as 1 minute. Flow conditions and wet well volume should be carefully examined to ensure the normal operating range of the wet well is not exceeded.

Equal-size pumps with two-speed motors can be controlled to produce an output comparable to adjustable-speed pumps.

Occasionally, two different sized constant-speed pumps are controlled to produce similar results. Consider the case of three pumps, each with single-speed motors. One has a nominal capacity of 1Q, and two have nominal capacities of 2Q. As level rises, pumps can be started and stopped to produce pumping station nominal outputs of 1Q, 2Q, 3Q, 4Q, and 5Q. Wet well volume should be analyzed to ensure pump-starts-per-hour criteria are not exceeded.

8.2 All Variable-Speed Pumps

The simplest case of variable-speed pumping stations is one active pump. Two setpoints (high and low) will dictate start and stop of the pump. Pump speed typically will ramp up from minimum to maximum speed as the level rises from low to high. An equivalent control scheme uses proportional mode feedback control with preset gain and initialization factors. The minimum speed setting must be set to keep pump operation within the pump operating curve, and thus is determined by the system head-capacity curve. For centrifugal pumps, this minimum speed setting is typically 70% to 80%.

Proportional integral (two-mode) control of a level setpoint has been attempted but, in general, is not recommended. Although successful in certain applications, this scheme has potential problems. If tuned too tightly and if flow varies rapidly, pump speed will oscillate wildly in attempting to maintain the setpoint level.

For multiple, identical, variable-speed pumps, each pump starts and stops via unique setpoints. A less common alternative starts and stops multiple pumps via a total of only two setpoints used in conjunction with time delay logic. A common practice is to operate all pumps at the same speed in a load share mode, which maintains the same capacity for each pump at the system head. As the next pump starts, the speed of all running pumps is reduced from maximum to some lower speed (not necessarily minimum pump speed). For instance, in a two-pump case, the lower speed is selected such that the flow discharge of the two pumps is slightly less than the output of one pump at maximum speed. This provides an overlap, preventing possible excessive cycling between the one pump on at maximum speed versus two pumps on at a lower speed.

Although the load share mode is common, algorithms sometimes are developed so that two pumps run at different speeds, that is, staggered arrangement. A typical reason for such algorithms is to permit the use of a common variable-speed drive for two pumps (Gottliebson, 1977).

The speed at which each variable-speed pump runs typically is determined by the ramping curves that are part of the overall control algorithm. Each ramping curve typically provides pump speed as a function of wet well level. For multiple pumps operating in a load-share mode, there may be a single or multiple ramping curves.

If there are multiple ramping curves, the ramping curve followed at any instant typically depends on wet well level or the number of pumps on or both. Multiple ramping curves can each occupy separate vertical segments of the wet well or the vertical segments of the curves can overlap. The latter practice makes effective use of wet well volume and is especially useful if the designer wants to reduce wet well depth.

Figure 7.12 presents the control sequence for the three active-pump, variable-speed pumping system shown in Figure 4.24. The pumps run in a load-share mode. Only one ramping curve is used, though often multiple ramping curves are used instead.

8.3 Combination of Constant- and Variable-Speed Pumps

Particular care is required in controlling stations having a combination of variable- and constant-speed pumps. It is important to keep the variable-speed pump within its operating envelope. A careful analysis is required, in part because the total dynamic head (TDH) of the variable-speed pump changes with pump speed, which is related to the wet well level.

Consider a pumping scheme with constant- and variable-speed pumps. At certain wet well levels, the variable-speed pump may be at or near shutoff head because of the constant-speed pump. Possible damage to the lag pump from radial thrust forces may result. A possible solution narrows the speed range of the variable-speed pump so its range of TDH is compatible with that of the constant-speed pump. However, narrowing the speed range

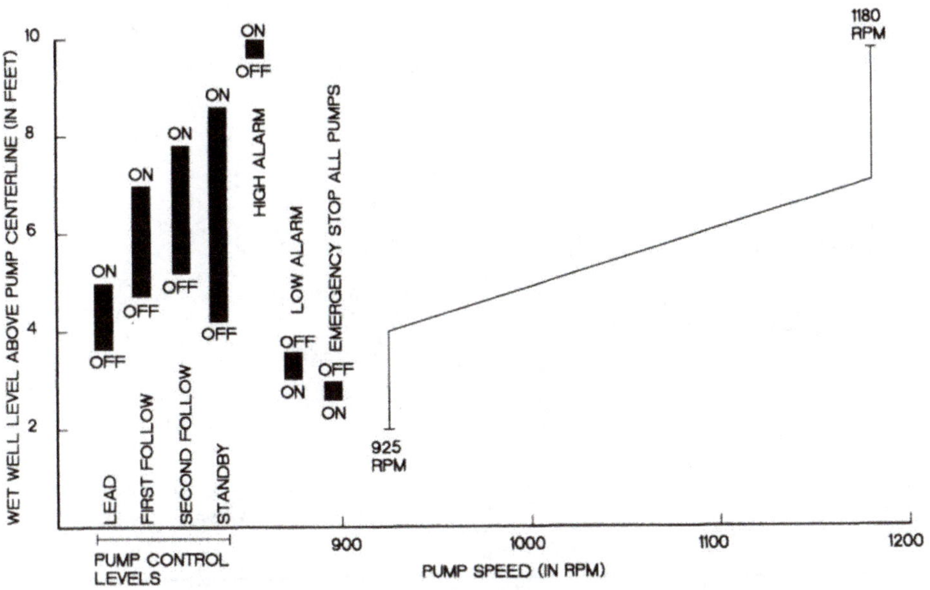

FIGURE 7.12 Control Algorithm for Pumping System of Figure 4.24 (ft × 0.3048 = m)

reduces pumping station flow range, which may be unacceptable from a process standpoint.

Figure 7.13 and Figure 7.14 present example control algorithms of a pumping station comprised of both constant- and variable-speed pumps.

The Figure 7.13 algorithm has no discontinuities or "gaps" in the pumping station flow output as level rises in the wet well. The pumping station consists of one lead 63-L/s (1000-gpm) variable-speed pump and one lag 32-L/s (500-gpm) constant-speed pump. The variable speed operating range is 3.33 to 1, or 20 to 63 L/s (300 to 1000 gpm).

As level rises in the wet well, the lead variable-speed pump eventually reaches 63 L/s (1000 gpm). As level continues to rise, the 32-L/s (500-gpm) lag pump starts. Once the lag pump starts, the lead variable-speed pump reduces or "ramps down" its speed so its output is also reduced to 32 L/s. Total pumping station output remains at 63 L/s, assuming a slow rise. There is no flow discontinuity or "gap." Control algorithms such as this generally have few problems associated with them.

The Figure 7.14 algorithm exhibits a flow discontinuity or "gap" as level rises in the wet well. The pumping station consists of one lead 63-L/s (1000-gpm) variable-speed pump and one lag 63-L/s (1000-gpm) constant-speed pump. The variable-speed pump range is 3.33 to 1, or 20 to 63 L/s (300 to 1000 gpm).

As level rises in the wet well, the lead variable-speed pump eventually reaches 63 L/s (1000 gpm). As level continues to rise, the 63-L/s lag pump

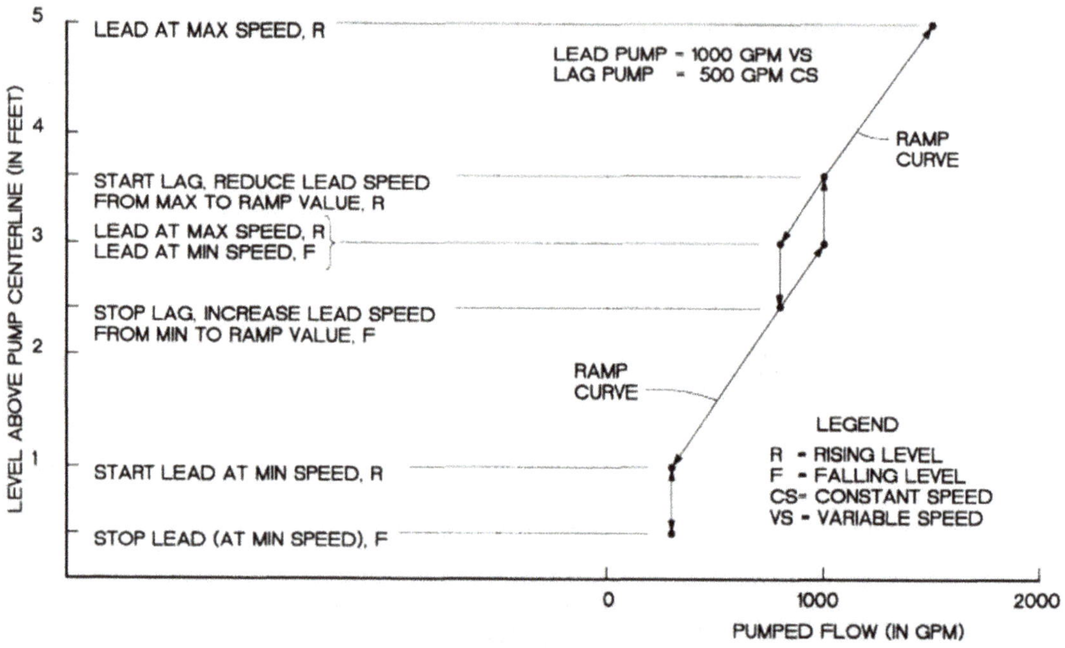

FIGURE 7.13 No Gaps in Control Algorithm—Combination of Constant- and Variable-Speed Pumps (ft × 0.3048 = m and gpm × 6.308×10^{-5} = m^3/s)

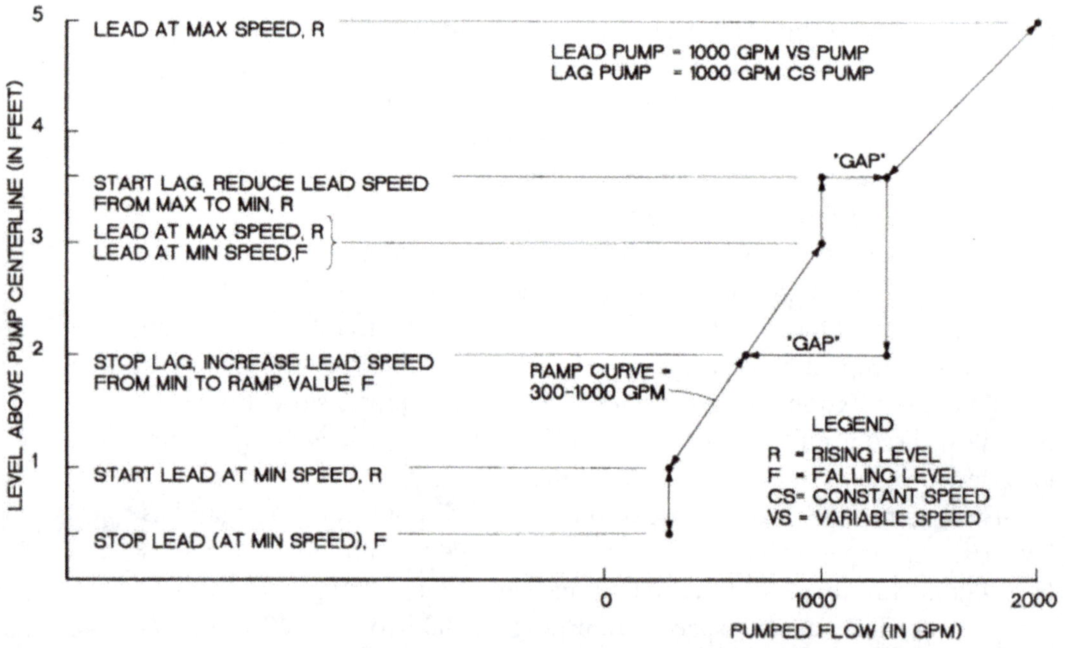

FIGURE 7.14 Gaps in Control Algorithm—Combination of Constant- and Variable-Speed Pumps (ft × 0.3048 = m and gpm × 6.308×10^{-5} = m^3/s)

starts. The lead pump is "ramped down." However, at its minimum speed, its output is not zero, but 20 L/s (300 gpm). Thus, the pumping station output jumps from 63 to 83 L/s (1000 to 1300 gpm) for an incremental change in level, and the pumping station has a 20-L/s "gap."

If inflow to the pumping station remains at 63 L/s (1000 gpm), the pumping station will rapidly pump down and the lag constant-speed pump will soon stop. This scenario promotes possible excessive cycling for the installed motor horsepower.

Gaps typically occur when there is insufficient variable-speed pumping capacity during a constant-speed pumping transition. Although gaps in such algorithms can cause significant cycling problems, some practitioners successfully use them. As long as "gaps" can be tolerated, equal-sized variable- and constant-speed pumps can be used, which is an advantage. Such advantages might include interchangeable parts, ease of wet well layout, and ease in pump selection.

A 1135-L/s (18 000-gpm) peak flow pumping station that has "gaps" in its algorithm has been in successful operation for more than 8 years (Bergs, 1993). The pumping station consists of four equal-sized pumps, with two constant-speed pumps and two variable-speed pumps. Of the two variable-speed pumps, one is active and the other is a standby. Gaps do not appear to pose a problem for several reasons. The wet well is large enough to keep pump cycling time within recommended limits. Flow to the pumping station is either very low or very high. Thus, flows that would accentuate adverse "gap" behavior are infrequent, and the gaps themselves are not particularly large.

In summary, algorithms with "gaps" can produce both excessive cycling and large flow swings to the receiving entity. If "gap" algorithms are to be used, the pumping station should be designed with caution and thoroughness to reduce possible cycling problems.

8.4 Alternation

Algorithms are also developed to alternate pumps. Pump alternation exercises the pumps, resulting in evenly distributed pump wear. However, there is a school of thought that equal pump wear promotes simultaneous pump failure and should be avoided.

Various hardware exists to alternate and select lead and multiple lag pumps in pumping stations. These include electromechanical alternators, electromechanical stepping switches, microprocessor-based alternators, lead and lag selector switches for manual alternation, and manual alternation by plug and jack systems, which are typically reserved for use for sequences with many pumps. Programmable controllers can also be programmed to cause alternation.

Alternation schemes range from simple to complex. An example of a simple alternation scheme is to alternate the lead pump in a duplex pump station. Complex automatic alternation schemes include first on and first off, and last on and first off, with initiation by time clock. The first on and first off scheme turns off the pump with the longest run time during the previous cycle and starts the pump with the longest rest time. Another complex scheme provides uneven run time on pumps. Even run time concerns some practitioners in that pumps may all fail at the same time.

For pumping stations routinely visited by operators, the most cost-effective scheme is to operate the pumps in a fixed sequence periodically changed by the operator with a selector switch or jacks.

Common ways to automatically initiate alternation include reaching low level when all pumps are off, at periodic real-time intervals, or when elapsed time of a lead pump exceeds a preset value.

Sometimes, automatic alternation is used in the retrofit of an undersized existing wet well because alternation effectively increases the time between starts of an individual pump. For instance, consider a duplex constant-speed pumping station with a required firm pumping capacity of 1Q. Each pump has a capacity of 1Q. Starter and motor sizes are such that they require 30 minutes between pump starts.

At critical inflow (Q/2) to the existing wet well, the one firm pump might have a minimum time between starts of only 15 minutes. However, by alternating pumps, the time between starts of any individual pump is an acceptable 30 minutes. Because acceptable starts per hour require that two pumps must be available for operation, pump capacity is not a truly firm 1Q. To rigorously make this capacity firm, another 1Q pump must be installed. The designer must decide if the total installed cost of this arrangement is less than the cost of enlarging the wet well.

8.5 Abnormal Operating Conditions

Abnormal operating conditions often pose problems, which must not be overlooked in designing the control system. The control system should provide an orderly startup after a power outage.

To reduce hydraulic transients and simultaneous motor inrush currents, pump starts often are staggered by the use of appropriately set time-delay relays. In contrast, for some systems, hydraulic transients unexpectedly are greatly reduced when pumps start up simultaneously because pressures that slam check valves are balanced out. However, the electrical system would experience additive motor inrush currents.

Protective control circuits should be analyzed to ensure they enable the process equipment to automatically restart after an outage.

For control systems with programmable controllers, its processor and input/output peripherals should be capable of resuming operation after an outage or "brownout." Typically, this is feasible with some minor reconfiguration.

As previously stated, programmable controllers benefit from power conditioning. Some designers choose to use a UPS to provide conditioned power. Others view the UPS as unnecessary because the programmable controller is only of use with main power available to the pumps. A UPS will reduce problems associated with "brownouts." However, if a UPS is used, the control algorithm must be carefully reviewed to ensure proper process equipment operation after a power outage. For example, the UPS-backed programmable controller may sense a false fail of process equipment during a power outage. Once power is resumed, the "failed" equipment might erroneously be interlocked from restarting.

Often for its own protection, a pump is shut down on an abnormal low level and provisions must be made to restart the pump, though manual reset generally is not recommended. An appropriate method is to restart the pump automatically as level rises to some higher level. This introduces deadband, which reduces excessive pump cycling.

Excessive pump cycling can prevent proper operation of a pumping station. In one instance during low inflow, a remote pump was forced to shut down because of high motor temperature developed by excessive cycling on an abnormally low level. Once inflow resumed, the station overflowed because the pump could not restart until manually reset.

9.0 VARIABLE-SPEED DRIVES

Variable-speed pump drive systems are also used for reasons other than smoothing flows to a receiving point. Variable-speed pump drive systems also reduce the number of motor starts, which increases motor life and reduces power demand charges. The number of motor starts also reduces transients to the pumping and piping system, which extends mechanical equipment life.

Variable-speed pump drive systems may also offer potential energy savings, but this is typically more of a factor in higher pressure water distribution systems. Potential energy savings are more achievable in systems with more frictional losses relative to static lift.

Because required wet well volume is less, variable-speed pump drive systems and their algorithms are also used in either retrofitting an undersized wet well or constructing a wet well from scratch to save construction costs. Variable-speed drives are covered in more detail in Chapter 6.

10.0 MISCELLANEOUS

10.1 Installation

Proper installation of instrumentation and other control components sometimes is overlooked. Fortunately, several firms are starting to specialize in supplying installation accessories, which should facilitate good installation practices. These commercially available accessories include such items as insulated enclosures and heating systems for field instruments to winterize them, modular design instrument supports, sunshades, and pre-insulated (and traced) tubing bundles.

Recommendations of suppliers should be followed, but sometimes this is not completely possible. For instance, the preference for installing most flow meters is vertical with upward flow, but mechanical layouts often preclude this. Required straight upstream diameters for some flow meters can be substantial. At least one manufacturer of Venturi-type flow meters accommodates less than ideal conditions by providing the amount of accuracy loss if the amount of straight upstream pipe is less than recommended.

Structural support for instruments and control panels should resist corrosion. Stainless steel (316 preferred) or aluminum members are recommended, as are stainless steel fasteners.

Because of their weight, control panel support should be reviewed closely with wind force considered. The temptation to handrail-mount instruments and control panels should be resisted. Though sometimes overlooked, the U.S. Occupational Safety and Health Administration regulations require the top handrail be unobstructed, increasing installation complexity. Especially if rain hoods are provided, the wind force can damage the handrail. Hanging components on handrails also reduces free clearance.

Because of its corrosion resistance, stainless steel tubing for applications such as between differential pressure primary elements and transmitters is probably worth the extra expense over copper. In cold areas, tubing should be heat traced or provisions made for water purges. Solar shades should be provided for instruments in direct sunlight with temperature-sensitive components (i.e., most solid-state devices).

In areas with wide temperature swings of moist air, condensation in control panels and conduits can occur. Panels should be equipped with panel drains and breathers. Condensation in conduit containing the cable between the primary element and its transmitter can cause problems, though sealing conduits with caulk might help. Figure 7.15 details how the condensate can be allowed to drain while still maintaining the watertight integrity of the conduit.

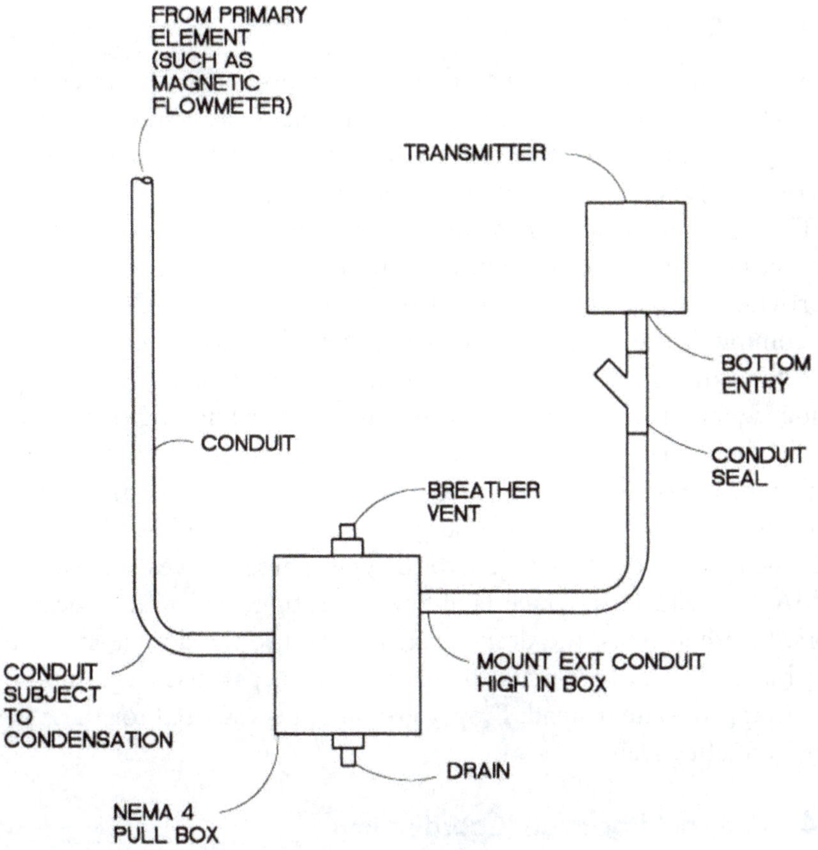

FIGURE 7.15 Conduit Condensate Drain Detail

10.2 Seal Water

Several types of centrifugal pumps require seal water; there are several seal water control systems to consider. For a pumping station in water-rich areas with relatively inexpensive pumps, continuous-seal water flow even with the pumps off might be considered. However, in the opinion of some practitioners, this practice should be avoided in that water should always be conserved.

The seal water solenoid valve often is interlocked with the pump control circuit to supply water only when the pump is on. Fail open (i.e., deenergized to open) solenoids are preferred because seal water will be supplied to the solenoid even if power to the solenoid is lost. For expensive pumps, monitoring seal water by pressure switch and subsequent pump shutdown may also be warranted.

10.3 Surge Suppression

Where lightning strikes are possible, surge suppression on system components should be considered. In lightning-prone areas, surge suppression is strongly recommended. Even in areas that are not lightning prone, some experienced practitioners recommend surge suppression.

The designer must exercise judgment in determining which equipment needs surge suppression. In lightning-prone areas, surge suppression should be provided for alternating current power to control panels and transmitters, communication lines (telephone lines for telemetry systems and antenna cable for radio and microwave transmission systems), and control and analog signal wiring to devices located in another building or outdoors.

The designer must also decide on the type of surge suppressor. In lightning-prone areas, alternating current power and analog signal lines should be furnished with fast-response surge suppressors with low clamping voltages, such as the double-stage hybrid type. These devices are available in NEMA 4X enclosures, which facilitate protecting transmitters located outdoors. In other areas, the designer again can use the double-stage hybrid type, but the lower cost metal oxide varistor or gas discharge suppressors may be appropriate instead. Proper grounding is essential for these devices to operate effectively.

10.4 Contract Document Coordination

The I&C system is integrated in the sense that several subcontractors or vendors have components that either are part of or are controlled by the I&C system. Any of these weak links can break the I&C system "chain."

Coordination problems during construction are common, but these coordination problems can be reduced by using judgment in assigning I&C components to a particular section of the specifications.

In large, complicated pumping stations, project execution often is assigned to an I&C systems integrator. Such integrators provide a variety of services, including field instrumentation purchase and field calibration, fabrication of control panels or the equivalent, programming and configuration services, and testing. They typically have trained field personnel who often provide on-the-spot troubleshooting services. They also produce thorough and well-organized documentation of the I&C system, which is of great value.

Pump vendors sometimes provide control systems as part of their standard package. These control systems are either done inhouse or done by contract by another firm. Control requirements of the project should be checked against the logic provided as part of the standard package. If much different, an I&C systems integrator should be considered. Sample documentation of previous package control systems should be reviewed during design to check for quality. If quality appears to be a problem, an I&C systems integrator should again be considered.

Smaller projects sometimes do not justify the use of an I&C systems integrator, but rather the control system is provided by the pump vendor or even by the electrical subcontractor. Some field instrumentation still is required for these projects. If instrumentation is minor such as just a few float switches, they can be furnished either by the pump vendor or the electrical subcontractor. More care should be given for more complicated components such as Venturi tubes. Whatever section such an item is specified in, the specification should include requirements for field services, including instrument calibration and field testing.

The I&C system is closely related to the electrical subsystem, and the specifications should be coordinated so that adequate documentation is required by subcontractors to facilitate construction. Panel elementary drawings that include terminal numbers must be provided. Interconnecting drawings must be provided that show wiring (including intermediate terminal junction boxes) between control panels and motor control centers.

10.5 Training and Maintenance

Many control systems of pumping stations do not function well after a time because of poor maintenance. Field instruments should be periodically calibrated, inspected, and repaired. Hardware and software control system problems should be diagnosed and corrected.

Maintenance can be done inhouse or a maintenance contract can be procured with a systems integrator. If done inhouse, instrument technicians, electrical technicians, or trained individuals with an interest in control systems can perform the required maintenance and diagnosis. Some, but not all, of these skills can be learned on the job. Formal training courses are beneficial.

Training courses include topics such as instrument calibration and programming, and the design of large projects often include formal training of some operations personnel. Ideally, the owner should make I&C training an ongoing program.

11.0 TELEMETRY SYSTEMS

Monitoring of remote pumping stations is often done at a central facility, and sometimes central facility control of the pumping stations is also provided. Telemetry systems link remote pumping stations and the central facility. Regardless of the telemetry system used, the design of the central facility control system should provide for any expected future expansion.

11.1 User Needs

Central facility monitoring keeps personnel aware of operating conditions at remote pumping stations. Such knowledge should reduce the number and quantity of spills by providing a quicker emergency response. Noncritical abnormal operating conditions can be ascertained and corrected before a critical condition develops.

Though the above reasons are not directly related to cost savings, for some owners they offer enough justification for a telemetry system.

Some telemetry systems are justified on a savings of operations costs, such as energy savings and the labor savings that result from reduced trips to inspect and acquire data from remote pumping stations. Telemetry systems are particularly justified where overflow reporting is required within a set time limit by a regulatory permit. A single fine can easily approach the cost of a telemetry system (Peterschmidt, 1993).

As a minimum, several key abnormal conditions should be monitored from each pumping station. These include abnormal high levels, a lost communication link between the central facility and the remote pump station, and a loss of power to the remote pumping station (Weirens, 1992). Another condition always worth monitoring is a general trouble alarm from the remote pumping station. Besides monitoring, the central facility can also be configured to control these remote pumping stations.

In stormwater and wastewater system, central facility control is required when systemwide management is practiced (e.g., flow storage or routing to different WRRFs for peak shaving or equalization). Sometimes central facility control is used to increase reliability of remote pumping stations. Thus, if local control at the pumping station malfunctions, the central facility assumes control. For such cases, the control systems must be carefully examined to ensure central facility override control is independent of local control.

11.2 System Types

Telemetry system types include automatic telephone dialers and tone and digital telemetry. Supervisory control and data acquisition systems can be built using many telemetry system types. However, digital telemetry systems lend themselves to complex SCADA functions such as archiving data. Current SCADA systems of at least moderate complexity invariably are digitally based.

Automatic telephone dialer systems are often used in smaller systems. The automatic dialer uses a dial-up line, which is an advantage because of its low cost. On an alarm condition such as a high wet well level, the automatic dialer calls a preprogrammed number with a voice message. Calls are placed to such locales as the central facility, an answering service, a pager assigned to an "on-call" employee, or directly to an "on-call" employee. Most dialers have battery backup so they can function during a power outage, but systems with multiple dialers can experience problems with common systemwide alarms (e.g., a power failure).

Automatic dialer capabilities include messages that give process variable information, pump run time, station totalized flow, and even some SCADA functions. These include starting and stopping pumps and printing out at the central facility all abnormal conditions for a pumping station in alarm.

Some dialers have had problems placing long distance calls. Lightning surges have also caused dialers to issue false alarms (Ouzts, 1992). The dialer itself should be protected with a surge suppressor.

Some existing pumping stations include auto dialers that are connected with a phone line, but some installations use a cellular connection. Multiple cellular systems include additional features that can allow for the operator to also view the station alarms and status remotely from a secure password-protected website. A cellular system can sometimes be a lower cost solution for remote locations where it may be more costly to bring in a phone line, but where cell service is available.

Tone telemetry conveys information over a communication link by means of audio tones. The two basic types of tone telemetry include amplitude modulated and frequency shift keyed (FSK). A given amplitude modulated tone telemetry device uses only a single carrier frequency and thus conveys only two-state information (on–off). Thus, a nonmultiplexed amplitude modulated transmitter or receiver would have only 1 Form C contact. It is difficult to detect a loss of carrier with amplitude modulated tone telemetry.

Though more versatile and secure from noise than amplitude modulated tone telemetry, FSK tone telemetry is more expensive. A given FSK tone telemetry device uses a single-carrier center frequency and frequency drifted mark and space frequencies to convey information. Thus, a nonmultiplexed FSK channel can have up to 3 Form C contacts, one for each of the carrier, mark, and space frequencies. One of the contacts can be used to detect communications failure.

The communication link for tone telemetry is typically a leased telephone line. Both discrete and analog signals can be communicated. As a minimum, status and alarm discrete signals are sent from the pumping station to the central facility. In some systems, the central facility also sends supervisory control signals to pumping stations. Some systems also send analog signals to the central facility to provide pumping station flow and wet well level information.

Multiple analog and discrete signals from a single pumping station can be transmitted over a single telephone line using multiplexers that interface with the tone telemetry equipment. Multiplexing is done either by time division multiplexing or multiple tone channels (frequency division multiplexing). The time division multiplexers use digital techniques that add to the security of the data.

Because of their flexibility and low cost, digital telemetry systems are more common than tone telemetry systems in new installations. Accordingly, digital telemetry systems are integral to the design of SCADA systems.

11.3 Supervisory Control and Data Acquisition Systems

Supervisory control and data acquisition systems are used anywhere from small, straightforward pumping systems to large, complicated ones in which there is a desire for systemwide management. At the central facility, the SCADA system provides interactive displays for status monitoring and control, alarm processing, trending of analog variables such as flow, report generation, and historical data collection.

Most SCADA systems today are microprocessor-based, using personal computer systems. A typical, simple SCADA system includes a single master computer with a monitoring screen and a report and alarm printer. It should be cautioned that a single computer SCADA system often contains possible single points of failure.

Sophisticated SCADA systems might include multiple computers linked together on a local area network (LAN) or a "hot standby" configuration. Multiple computers are primarily provided for redundancy. There is some flexibility in assigning tasks to each computer, though even SCADA systems with multiple computers on a LAN can contain a single point of failure (Peterschmidt, 1993).

In some systems, a satellite facility at a different location augments the central master facility. A typical communication link between the master and the satellite would be a leased telephone line, radio, or fiber communications. Figure 7.16 presents a block diagram for a SCADA system.

The SCADA system communicates with the RTU of each individual pumping station by either of two basic methods: polling or contention. Some vendors use a combination of both. In polling, the central host periodically polls the pumping station RTUs to acquire data. In contention, RTUs only communicate with the central host if there is a change in state. A common problem with contention occurs if many RTUs detect a simultaneous failure, such as systemwide power failure. This can overload communications channels, and manual intervention is required for the system to recover (Peterschmidt, 1993).

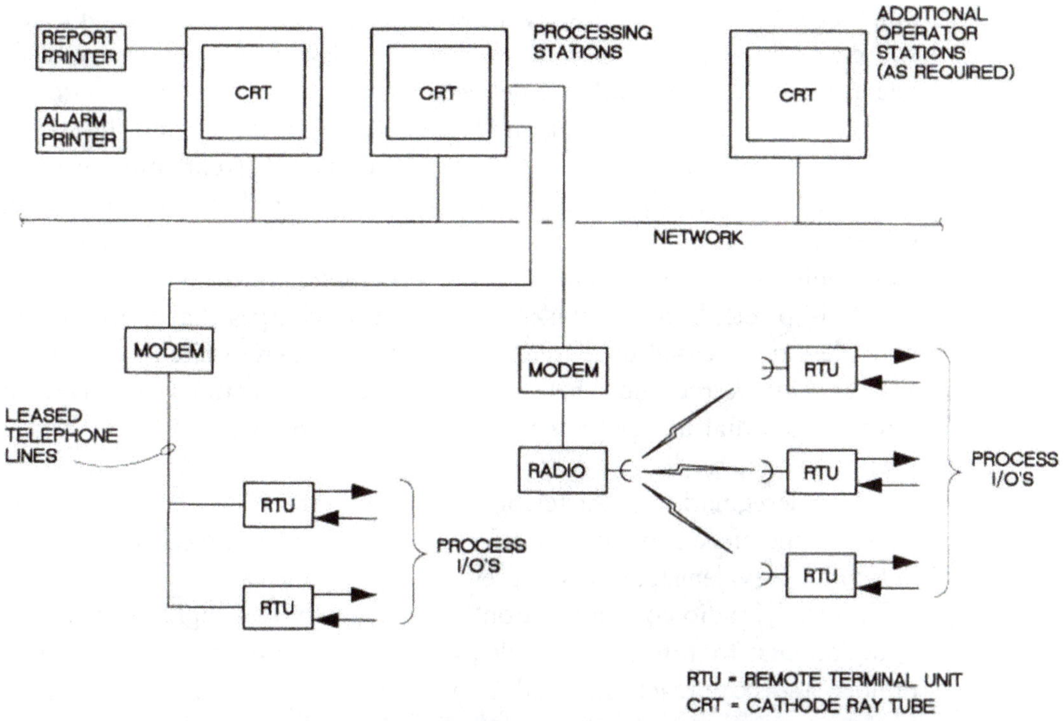

FIGURE 7.16 Supervisory Control and Data Acquisition System Block Diagram

In the design of SCADA systems, the appropriate polling time should be examined. Often, the initial desired polling time pushes the capabilities of equipment commercially available at a reasonable cost (Ouzts, 1992). A slower polling time may be more appropriate and realistic. Alternatively, a contention protocol or multiple common channels should be considered.

Even with SCADA systems, primary control typically resides at the pumping station itself, external to the SCADA system. The SCADA system could be designed to override the primary control mode, either by operator manual intervention or by altering the automatic strategy. For large sophisticated systems, logic would be distributed between the pumping station and the central facility. At the central facility, systemwide management decisions would be made. In some SCADA systems, primary control resides at the central facility. If a communication failure is sensed, control reverts to local automatic or manual control.

11.4 Communication Links

A variety of communication links are available that convey information between the central facility and remote pumping stations. Several possible links include radio channels, leased telephone circuits, leased metallic telephone pairs, dial-up telephone networks, cellular telephone systems, cable television systems (broadband network), microwave radio links, satellite channels, fiber optic networks, and leased fiber optic lines. Some links such as leased metallic telephone pairs (which use direct current continuity circuits) are less available as phone companies phase them out of service. Others are appropriate only under special circumstances. Two of the most commonly used are radio channels and fiber optic networks.

Dial-up telephone networks have monthly charges that are typically much less than leased telephone circuits. Thus, where limited amounts of data are transferred and where brief delays can be tolerated in conveying information, dial-up telephone networks might be preferred over leased telephone circuits.

With deregulation of the telephone system in the 1980s, leased telephne lines became more expensive on average, making the radio communication link more prevalent than it was previously.

Typically, radio communication link systems have a higher initial construction cost than telephone links, but lower operating costs. Typically, radio systems can reach the break-even point in 3 to 5 years. The telephone leasing cost is also subject to escalation. Because the owner has less control over the telephone link, there may be difficulties in obtaining timely maintenance and troubleshooting.

If considered for use, the radio communication link must be both technically feasible and available. Suitable radio frequencies are in short supply in many populous urban and suburban areas, where licenses needed to operate the radio communication link may not be available. The first priority is to determine if licensable frequencies are available.

Topography must be considered in determining technical feasibility. Flat areas where there is a line of sight between the central facility and remote pumping stations promote radio communication. Radio "repeating stations" are used as part of the radio communication link in more rugged areas, but add cost and complexity to the link.

In evaluating a radio communication link, a study phase is conducted to determine the radio system requirements and its economic feasibility. Available radio frequencies are determined, and a path analysis or propagation study is performed.

Several firms offer frequency-coordination services to determine available frequencies in the proposed areas of operation. Frequency bands commonly available are midband, highband, very high frequency (VHF), ultrahigh frequency, and 900 MHz. These bands each have their own technical advantages and disadvantages. For instance, VHF bands are subject to "skipping," where, under certain atmospheric conditions, the radio waves "skip" off atmospheric layers and can travel up to hundreds of miles. The bands are also subject to different regulatory constraints.

Based on these and other factors, a feasible available frequency can be selected for license. During the design phase, firms complete the necessary paperwork for the user to file for a Federal Communications Commission license. Upon awarding of the license, the user has 1 calendar year to get the system operational or the license is revoked. Extensions can be requested, though they are not routinely awarded in some popular frequency bands.

During the study phase, an initial propagation analysis should be performed to determine the parameters necessary for reliable operation. This analysis can be based on available line of site between nodes. Personal computer programs are available to perform path analysis based on latitude and longitude coordinates of proposed sites in conjunction with topographic databases.

Propagation calculations and field tests can be used to verify or reject marginal or questionable paths. *Propagation* is a term that describes how radio wave intensity varies in space. Reliable link communication is only possible when the strength of the received signal is easily distinguished from noise. Factors affecting propagation include the selected frequency; distance between sites; topography; obstructions, including foliage and buildings; antenna types and heights; cable lengths and types; and transmitter power and receiver sensitivity.

Fade margin is a measure of signal strength in excess of that required to just distinguish the correct signal. The expected fade margin is determined as part of the propagation study. Each link should have a positive fade margin (e.g., a value of 20 dB).

Typical problems encountered with radio communication links include momentary interferences from sunspots and thunderstorms. Long-term interferences include "skip" and "bleed over" from splinter frequencies.

Figure 7.17 shows a remote pumping station furnished with a radio antenna.

There are two main types of fiber optic cable: single mode and multi-mode fiber. Multi-mode fiber is typically used for cable distances that are less than 305 m (1000 ft). Single-mode fiber is used when connecting two facilities that are kilometers (miles) apart. The advantage of fiber is that it has a higher bandwidth, allowing it to transmit data, voice, and video. A designer can also create multiple networks where the SCADA system is one pair of fiber strands and would not be interrupted by the high data exchange that may be seen on the video surveillance network.

One of the advantages of installing a fiber optic system is that it is possible to expand and easily accommodate future changes by replacing the transmission devices and reusing the existing fiber optic cable. For significant

FIGURE 7.17 Remote Pumping Station With Radio Antenna

distances where radio is typically used, fiber can be a more expensive option than using radio. This is due to the cost of acquiring right-of-way for the route and the high installation cost of the raceway system (innerduct, pull boxes, etc.). Another disadvantage is that fiber cable is more susceptible to physical damage where it is routed outside the owner's facility.

11.5 Remote Terminal Units

The use of "smart" RTUs is becoming more widespread than the conventional RTU, the only purpose of which is to interface signals with the communication link. Because local control algorithms reside in a "smart" RTU, the design is simplified.

Programmable controllers furnished with appropriate modems can also be used as RTUs. This practice is common and cost-effective if the programmable controller is also providing the local "stand-alone" control.

Lower-cost alternatives to SCADA systems are pumping stations furnished with a type of "smart" RTU that only periodically (e.g., daily) communicates to the central facility by telephone dial up (Morey, 1992). Information conveyed includes pump starts, pump elapsed run times, cycle times, and combined flow. These systems can also call up with unscheduled reports on an alarm condition.

Regardless of the type of RTU used, it must be interfaced with the local control system. The actual operator interface is typically on the local control panel. Simple systems may not have a local control panel. Instead, the operator interface would be on the motor control center itself.

12.0 RELIABILITY

Reliability is paramount for stormwater and wastewater pumping stations used in critical applications. Accordingly, reliability should be incorporated to the design of their control systems. For instance, a significant wastewater pumping station that would, if it failed, surcharge sewers must be designed for reliable operation. Similarly, reliable operation is required for those stormwater pumping stations that, in the event of failure, would flood roadway underpasses, tunnels, basements, and low-lying towns protected by levees.

12.1 Level Sensing Elements

Alarm-sensing devices should be provided independent of normal operating sensing devices. For instance, floats can be provided to alarm abnormal levels if the bubbler system used in normal operation fails.

For level-sensing systems used in normal operation, dual-sensing devices and transmitters could be provided, such as two bubbler systems or ultrasonic meters provided to measure wet well level. Bubbler systems could also be provided with dual compressors.

12.2 Control Hardware and Algorithms

The normal mode control algorithm can be supplemented by backup algorithms such as starting one or more pumps on an abnormal high level.

In pumping stations with both constant- and variable-speed pumps, often only the constant-speed pumps are configured to start on the abnormal high level. To plan for this, the designer provides enough wet well volume to prevent excessive pump starts and often includes the surcharged influent pipe as usable volume.

The heart of the control system could be provided with redundancy. For example, the main PLC might be backed up with a "hot" PLC. Also, a SLDC or a group of relays may back up critical parts of the control algorithm resident in the PLC. Transfer of control from the programmable controller could be automatic by means of a watchdog timer, which senses if the programmable controller fails.

For SCADA systems, manual override operation is sometimes performed from the central facility. Also, for some SCADA systems, dual communication links such as radio and leased telephone lines may be provided. Some "smart" RTUs now have backup dial-up telephone capabilities. If the "smart" RTU senses no poll, it dials up the central facility by exception reporting.

12.3 Control Power

The control panel could be designed to accept control power from two different sources. The system can be designed so that backup algorithms are powered directly from the motor control center control circuit of each individual piece of process equipment. Thus, these backup algorithms still would function if the control panel were to lose power.

Critical pumping stations are often provided with backup power sources such as a diesel generator. By design, power to all pumps may not be provided while under a backup power source. If control panel power happens to be tied into the bus that feeds the nonpowered pumps, the control algorithm would not function.

12.4 Pumps and Drives

The type of pump used should be examined. Because screw pumps can pump dry without incurring harm, their backup control algorithms are inherently simpler than those for centrifugal pumps. For instance, the programmable controller can be configured to fail such that the pump command-to-run contacts close on power or controller fail, causing the screw pumps to run continuously.

Some variable-frequency pump drives are available with full-voltage starters so the pump drive can be run at constant speed. The switch over to the full-voltage starter can be accomplished either manually by means of a selector switch or automatically upon failure of the VFD.

13.0 TESTING

For all but the simplest pumping stations, a formal test program starting at the factory is suggested for several reasons. The I&C system is comprised of customized, sometimes complex, logic. Misinterpretation and incorrect implementation of the design intent is possible.

Parts of the design intent may also be flawed. Many aspects of the control algorithm deal with high pumping and abnormal operating conditions, which might not be encountered by the pumping station for years unless a specific test is conducted.

Finally, for many projects, the contractor is rushing to meet contractual deadlines. Often, there is only a narrow window to commission the I&C system because it depends on completion of other trades (such as mechanical and electrical).

A formal testing program lends an air of rationality under these circumstances and reduces the tendency to assume the I&C system is correctly functioning.

At least four types of tests are appropriate throughout construction, including unwitnessed factory tests, witnessed factory tests, field readiness tests, and witnessed field tests. For complicated pumping stations, all four types are suggested. Unwitnessed factory tests rely on the good faith of the control systems supplier to develop and conduct appropriate tests. In an unwitnessed test, the control panel or equivalent is tested for proper wiring and testing algorithms.

In a witnessed factory test, the owner or the owner's agent (typically the design engineer) witnesses control system testing, which is conducted by the control systems supplier. Normally, the control panels or equivalent are inspected, though the bulk of the testing covers the control algorithms. Test procedures are developed in advance of the test itself and are often detailed. Testing typically involves the simulation of process conditions, and deficiencies are noted and corrected.

Field readiness tests do not test logic functions. Instead, these tests verify that I&C components are installed and working, with wires properly terminated and setpoints set correctly. These tests are typically unwitnessed but must be documented to serve as a basis to proceed with the witnessed field tests.

Witnessed field testing is important because it demonstrates that the I&C system accomplishes all functions required by the design. Accordingly, passing these tests is typically linked to contractual milestones such as substantial completion and final acceptance. Testing is similar to the witnessed factory tests, except that real process variables are used as much as possible. For instance, tests may include wet well pump downs of water or wastewater to demonstrate proper function of the lead and lag pumps. Any deficiencies are noted and retested (either witnessed or unwitnessed).

Witnessed field testing is often conducted as other interrelated systems are completed. Thus, problems with these interrelated systems will be identified as testing of the I&C system proceeds.

To reduce testing bottlenecks, it is best to have certain trades either at the test or readily available. Two common beneficial trades would be electrical and mechanical. If too many deficiencies are identified, testing progress becomes so slow that the test will need to be postponed until the interrelated systems are made operable.

14.0 REFERENCES

American National Standards Institute/National Electrical Manufacturers Association. (2020). *Enclosures for electrical equipment* (ANSI/NEMA 250-2020).

Bergs, A. (1993). Personal communication. R.V. Anderson Associates.

Briggs, B. (1993). Personal communication. Primary Flow Signal.

Cooper, R. W. (1992). Personal communication. ADS Equipment.

deGlanville, B. (1992). Personal communication. CH2M Hill.

DiNapoli, L. (1993). Personal communication. BIF.

Gemin, J. (1993). Personal communication. Proctor and Redfern.

Gottliebson, M. (1977). Explore the use of variable speed water booster pumps/2. *Water Wastes Engineering, 14*(10), 58.

Guagno, J. (1992). Personal communication. EMA Services.

Gunby, F. M., Jr. (1993). Personal communication. Metcalf & Eddy.

International Society of Automation. (2009). *Instrumentation symbols and identification* (ANSI/ISA S5.1).

Liptak, B. G., & Venczel, K., Eds. (1982). *Instrument engineers' handbook—Process measurement*. Chilton Book Company.

Morey, E. (1992). Personal communication. Atlanta, GA.

National Fire Protection Association. (2020). *Recommended practice for fire protection in wastewater treatment plants* (NFPA 820).

Ouzts, J. (1992). Personal communication. Turbitrol Company.

Peterschmidt, W. (1993). Personal communication. CH2M Hill.

Robinson, C. (1986). Obstructionless flowmeters: Smooth sailing for some, rough passage for others. *InTech, 33*(12), 33.

Water Environment Federation, American Society of Civil Engineers, & Environmental and Water Resources Institute. (2010). *Design of municipal wastewater treatment plants* (5th ed.; WEF Manual of Practice No. 8/ ASCE Manuals and Reports on Engineering Practice No. 76). Water Environment Federation.

Weirens, D. J. (1992). Personal communication. Consolidated Electric.

Whitenack, W. (1992). Personal communication. Town of Morristown, N.J.

15.0 SUGGESTED READINGS

Dulas, J. H., & Tisdale, J. C. (1992). Improve flow control with electronic drives. *Chemical Engineering Progress, 88*(2), 40.

Institute of Electrical and Electronics Engineers. (2014). *IEEE recommended practices and requirements for harmonic control in electric power systems* (IEEE Standard 519).

Instrument Society of America. (1986). *Standards and practices for instrumentation* (8th ed.).

International Society of Automation. (2003). ISA recommended practice for wiring methods for hazardous (classified) locations instrumentation part 1: Intrinsic safety (ISA-RP12.06.01).

International Society of Automation. (2013). *ISA definitions and information pertaining to electrical equipment in hazardous (classified) locations* (ANSA/ISA-12.01.01).

International Society of Automation. (2015). *ISA human machine interfaces for process automation systems* (ANSI/ISA-101.01-2015).

Jones, C. T., & Bryan, L. A. (1983). *Programmable controllers—Concepts and applications.* International Programmable Controls.

Jones, G. M., & Sanks, R. L. (2008). *Pumping station design* (3rd ed.). Butterworth-Heinemann.

Keskar, P. Y. (1988). PLC system with emergency generators and variable frequency drives: A hard knocks education. *Programmable Controls,* 7(4), 36.

Occupational Safety and Health Administration. (1983). OSHA safety and health standards [29 CFR 1910, OSHA 2206, 1910.23(e)].

Skrenter, R. G. (1987). *Instrumentation handbook for water and wastewater treatment plants.* Lewis Publishers.

U.S. Department of the Interior. (1984). *Water measurement manual.* U.S. Department of the Interior, Bureau of Reclamation.

Water Pollution Control Federation. (1984). *Process instrumentation and control systems* (Manual of Practice No. OM-6).

8

Appurtenances

1.0 SCREENING

The screening of wastewater and stormwater is typically used to remove large floating and submerged objects that may otherwise damage pumps and other equipment, obstruct pipelines, and interfere with the normal operation of pumping and any downstream treatment facilities. Screening that removes oversized solids is distinguished from comminution or grinding, which reduces the size of solids without removing the solids from the wastewater. A basic design decision is whether to screen only to the extent necessary to protect the pumping equipment or whether to maximize screenings removal. While many factors enter into the decision, screening to maximize solids removal is typically practiced only at water resource recovery facilities (WRRFs) and is discussed in detail in *Design of Water Resource Recovery Facilities* (Water Environment Federation [WEF], 2023). Generally, screening for wastewater and stormwater pumping stations is accomplished with coarse or bar screens.

Bar screens consist of vertical or inclined steel bars spaced at equal intervals across a channel through which stormwater or wastewater flows. Criteria used in their design include bar size, bar spacing, angle from the vertical, type of steel, channel width, approach velocity, and method of screenings removal. The most commonly used coarse screens are mechanically cleaned bar screens, although manually cleaned bar screens in smaller pumping stations, or those used with bypass channels, are not unusual. This section also addresses comminutors because of their use ahead of pumping equipment in some installations instead of bar screens.

1.1 Types of Screens

Coarse or bar screens, which are typically the first equipment encountered by influent flow before the pumping station, can include trash racks, manually and mechanically cleaned bar screens, and basket screens.

1.1.1 Trash Racks

Trash racks are bar screens with large openings of 40 to 150 mm (1.5 to 6 in.) that are designed to prevent logs, timbers, stumps, and other large, heavy debris from entering the pumping station. Trash racks, sometimes followed by bar screens with smaller openings, are principally used in stormwater and combined systems that carry high amounts of large debris. Trash racks may be mechanically or manually cleaned. Where space is limited, pumping stations sometimes have basket-type trash screens that are manually hoisted and cleaned.

1.1.2 Manually Cleaned Screens

Manually cleaned bar screens typically have 25- to 55-mm (1- to 2-in.) openings, with the bars set 30 to 45 deg from the vertical to facilitate cleaning (Figure 8.1). Screens are typically constructed of heavy, evenly spaced rectangular or round steel bars set in a channel. Stainless steel is typically used for wastewater applications, while carbon steel with hot-dipped galvanizing can be used for stormwater applications. Bars are held in a parallel, equal-spaced position by a toe plate across their lower ends and by horizontal spacing bars welded to the screen's downstream side. These horizontal spacing bars should be designed so that they will not interfere with raking the screen, and the lower bar should be above the normal flow area of the screen to prevent screening buildup at this cross connection. Manually cleaned bar screens typically are custom designed by the engineer and specially fabricated from drawings and specifications.

Although manually cleaned screens require little or no equipment maintenance, they may present significant labor-related maintenance costs. They demand frequent raking to avoid clogging or excessive system headloss and hydraulic grade line rise. Screens are manually raked onto a drainage area or perforated plate, where they drain before removal or disposal. Provision

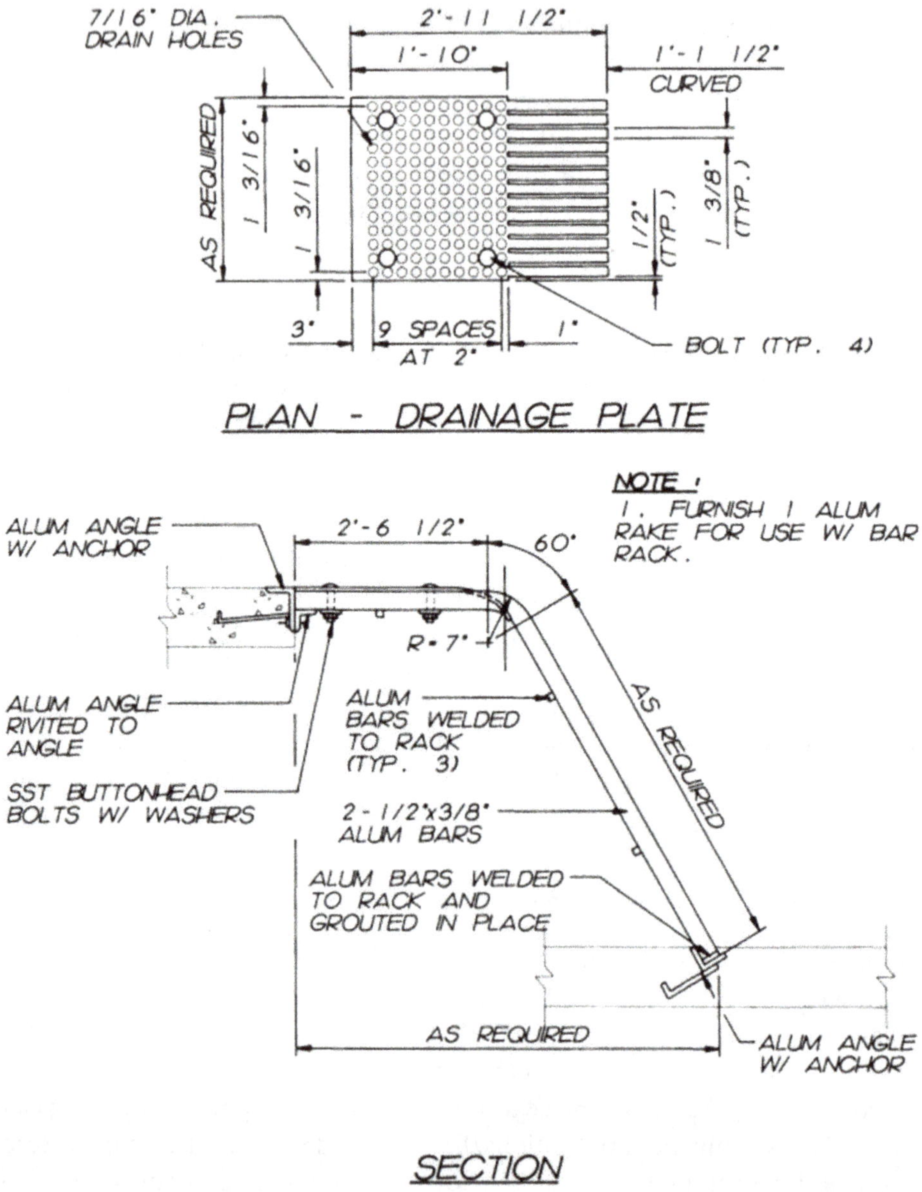

FIGURE 8.1 Manual Bar Screen Detail (in. × 25.4 = mm; ft × 0.3048 = m)

should be made for a raking tool that fits the screen bar spacing and a convenient method of handling the screenings to a storage container. Manual screens are typically limited to shallow channels of no more than 2 m (6 ft) so that the depth of reach does not pose a safety concern, such as leaning over the channel while removing a heavy screenings load. Removal of screenings may require a hoist or dolly. Manually cleaned screens exist primarily in older, small pumping facilities, and bypasses of comminutors and mechanically cleaned screens. Where entry to the wet well is required to rake the bar screen, confined space entry procedures and requirements should be followed. Because of the potentially intensive labor involved, the use of manually cleaned screens as the primary screening method should be avoided.

1.1.3 Basket Screens

Basket screens (see Figure 8.2) are typically in the form of a box with one partially open side facing the incoming sewer or are cylindrical with an

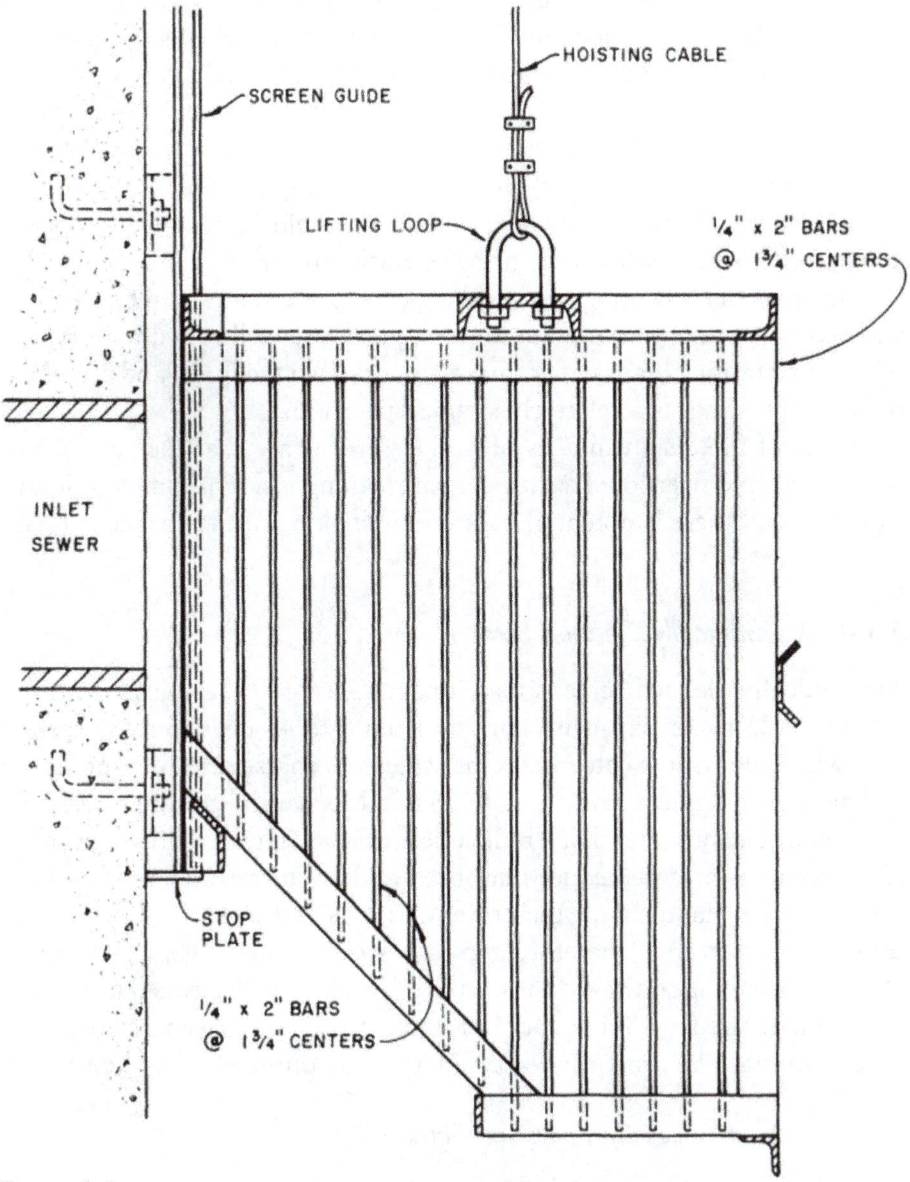

FIGURE 8.2 Basket Screen (in. × 25.4 = mm)

open top. If flooding of the screen is unlikely due to high wet well water level, the top of the screen box may be left open. Parallel flat bar construction of basket screens, particularly the bottom, makes cleaning easier than does round bar construction. Grating or mesh-type construction should not be used because such screens are harder to clean. Cross members for bar spacing and structural support should interfere as little as possible with basket cleaning.

Basket screens are typically raised for emptying and lowered by means of a corrosion-resistant cable and a winch, preferably one that is motor operated. The hoisting arrangement permits the raised basket screen to be maneuvered easily to a location for convenient removal of screenings to a storage container. Guides and supports should be designed to hold the screen in a stable position while it is being emptied. In the wet well, supports and guides should be installed so that the basket screen, when lowered, always returns to the proper position for effective screening of the discharge from the inlet sewer. Manhole steps or a ladder should be installed to permit access to the screen when it is in the operating position.

Some of the advantages of basket screens are that channel, dry well, and access facility space are not required; they are well suited for submersible pump stations that may not have any dry well facilities; and screenings removal (hoisting) is a relatively simple operation.

Some of the disadvantages of basket screens are that they are subject to clogging and overflow because screen cleaning is not automated (limited capacity) and there is potential hang-up or sticking of the basket in raising or lowering.

1.1.4 Mechanically Cleaned Screens

Mechanically cleaned screens have openings typically ranging from 5 to 40 mm (0.25 to 1.5 in.), with bars set from 0 to 45 deg from the vertical. This width of opening categorizes the screens as coarse screens. Fine screens will have openings less than 5 mm (0.25 in.). Mechanical cleaning, compared to manual cleaning, tends to reduce labor cost, improve flow conditions and screenings capture, reduce nuisances, and, in stormwater and combined systems, better handle large quantities of debris and screenings. A mechanically cleaned screen is, therefore, typically specified for larger (>3800-m^3/d [1-mgd]) pumping stations. Many types of mechanically cleaned bar screens are manufactured, including, but not limited to, chain-driven, reciprocating rake, rotating rake arm, gripper, catenary, and continuous bar screens.

1.1.4.1 Multi-Rake, Chain-Driven Screens

Mechanical bar screens with a chain mechanism (see Figure 8.3) to move the rake teeth through the screen openings are the oldest mechanized screening

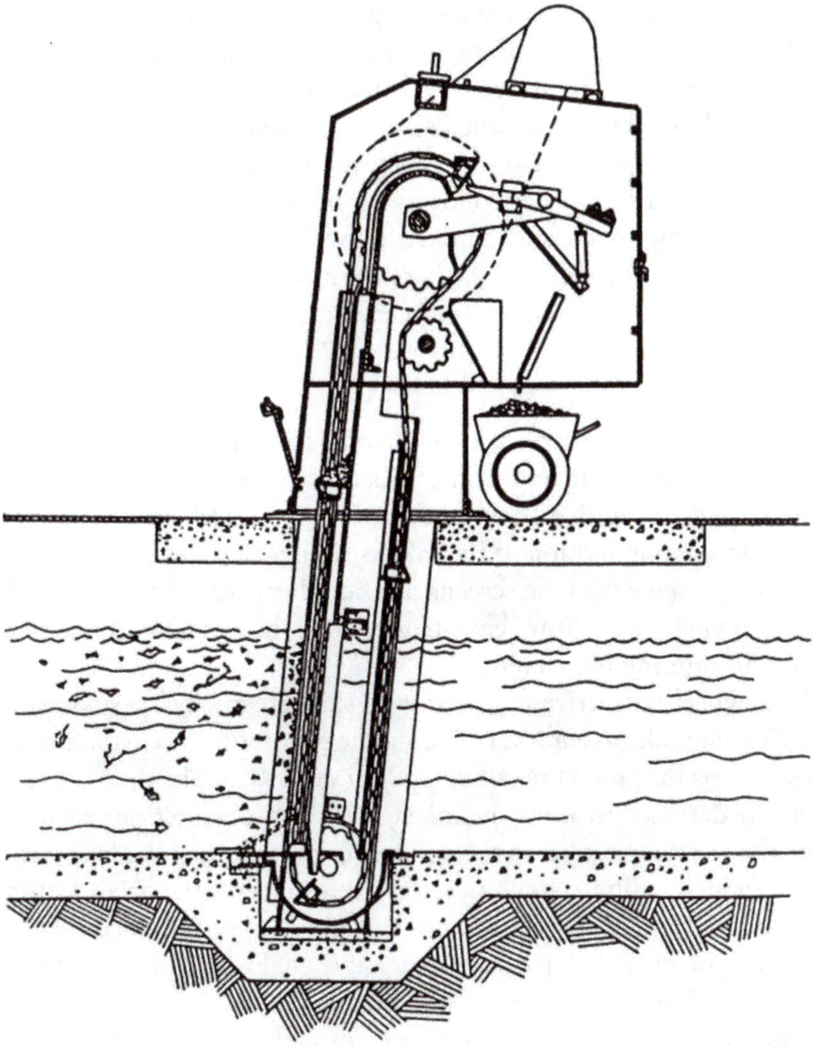

FIGURE 8.3 Chain- or Cable-Driven Mechanically Cleaned Bar Screen

devices. These types of screens are manufactured in several configurations: front clean/front return, front clean/rear return, and back (or through) clean/rear return. The front clean/front return type most efficiently retains captured screening by minimizing carryover. Such screens are used extensively in ordinary municipal wastewater treatment applications. A disadvantage of a rear return configuration is the possible return to the screened waste stream of debris that is not dislodged by the rake cleaning mechanism. For any of the front-cleaned configurations, bars are preferred to be trapezoidal in cross-section, providing a tapered opening to prevent wedging and trapping of solids between the bars. However, rectangular bars are sufficient for coarse screening applications. Disadvantages of front-cleaned screens include the possibility of bottom jamming by unusual deposits of trash, particularly if no trash racks are used. Also, the chain- or cable-driven raking mechanism consisting of submerged sprockets or other mechanical devices is subject to fouling by grit and rags. Periodic rake and operating mechanisms with the back- or through-cleaned screening configuration are protected by the bars from damage due to large objects in the wastewater stream. The back-cleaning action pushes solids away from the bars, preventing solids from being compressed and jammed between the bars. However, longer rake teeth protruding through the bars are more susceptible to bending and breakage than are the teeth of front-cleaned screens.

1.1.4.2 Reciprocating Rake Screens

The reciprocating rake screen, also known as a *climber screen*, is equipped with a front clean/front return mechanism (see Figure 8.4) that minimizes solids carryover or with a back-clean/back-return mechanism. The up-and-down reciprocating motion of the rake is similar to that of a person raking a manual bar screen. These screens are popular because their lack of submerged moving parts allows easy, trouble-free inspection and maintenance without dewatering the channel.

A cogwheel-type drive is paired with this screen. For these designs, the entire cleaning rake assembly, including the gear motor, is carriage mounted on cog wheels that travel on a fixed pin or gear rack. The drive mechanism is typically designed to allow the rake to ride over obstructions encountered during the cleaning stroke. In the unlikely event that the rake becomes completely jammed, a limit switch is activated to turn off the drive motor only after a predetermined number of cycles to clear the jam have been exhausted. Motors can be provided in fully submersible enclosures to allow continued operation during extreme peak flow events.

A disadvantage of reciprocating screens is the single rake, which limits the reciprocating screen's capacity to handle extreme loads, especially for deep applications in which cycle times are long. Also, these systems require high overhead clearance, which might limit their use in retrofits or require stair and access platforms to perform maintenance. Headroom requirements for reciprocating rake screens are greater than those for other types of screens. The headroom requirement can be determined by adding the length of the rake to the screening discharge height above the floor. Rake length is a function of the maximum water depth anticipated in the channel, plus some length to prevent submergence of the traveling rake motor.

1.1.4.3 Catenary Screens

The cleaning mechanism of a catenary screen (see Figure 8.5) consists of heavy tooth rakes held against the screen by the weight of its chain and driving force of the flow through the channel. The screen's name, *catenary*, stems

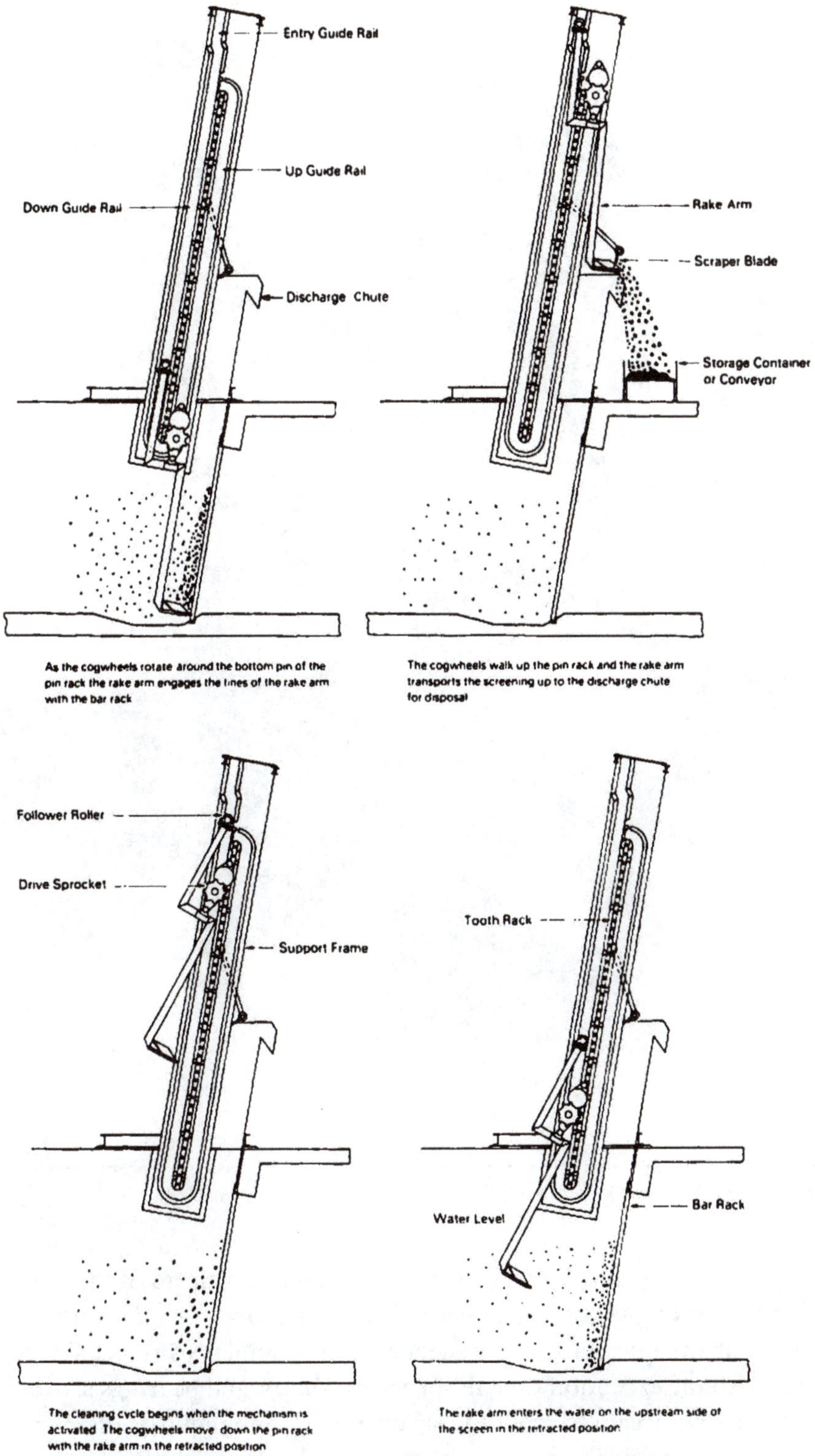

FIGURE 8.4 Mechanically Cleaned Reciprocating Rake Bar Screen

FIGURE 8.5 Catenary Bar Screen

from the catenary loop formed by the operating chain ahead of the screen. A curved transition piece at the base of the screen allows for removal of solids captured at the bottom. The catenary screen is similar to the chain-driven screen, with the exception that all sprockets, shafts, and bearings are located out of the flow stream to reduce wear, corrosion, and required maintenance. Rakes can be provided as stainless steel or polyethylene material.

Because the cleaning rake is held against the bars primarily by the weight of the chains, the catenary screen is recommended to be installed to a minimum 10-deg angle to provide the proper geometry for effective contact between the rakes and bar rack. Attention must be given to the lateral space requirements of the screen. Catenary screens offer low total height or headroom requirements similar to the chain-driven screen.

With the absence of a bottom sprocket, the rake can be pulled over large rags or solids that may be jammed in between the bars. This ride-over feature can prevent a jam alarm during operation but may cause premature wear on the rake and chain due to jarring movements when the rake settles back onto the bar rack. In the area of the jam, the rake will not make effective contact with the bar rack, leaving wide sections of the bar rack unraked until the jam can be cleared.

1.1.4.4 Continuous Self-Cleaning Screens

Continuous self-cleaning screens (see Figure 8.6) consist of a continuous "belt" of acrylonitrile butadiene styrene or other synthetic material elements

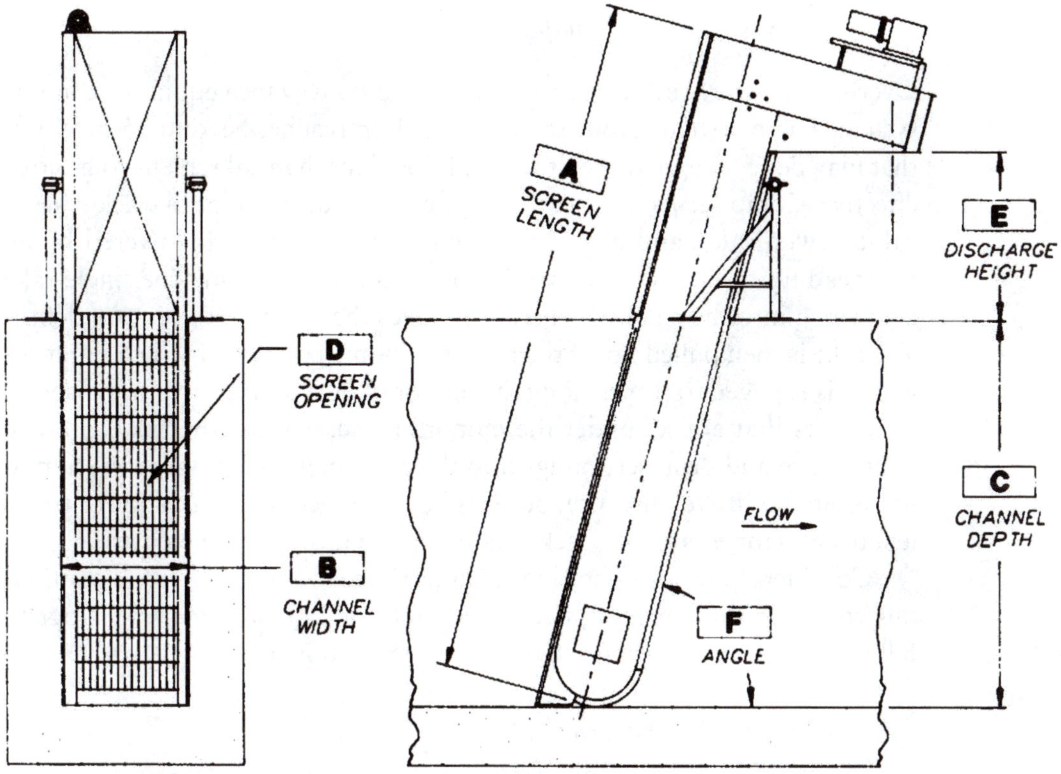

A = 8' to 25'
B = 12" to 120"
C = 2' to 16'
D = 3/16" to 3/4"
E = 3' to 5'
F = 60 to 80 degrees

FIGURE 8.6 Continuous Self-Cleaning Screen (in. × 25.4 = mm; ft × 0.304 8 = m)

that are pulled through the wastewater to clean the entire length of the screen. Screen openings are designed with both horizontal and vertical limiting dimensions; the vertical spacing is slightly larger than the horizontal spacing. The greater solids-handling capacity of these screens allows smaller openings to be used with greater capture of solids from the waste stream. Also, the continuous screening action of these units allows for efficient removal of large quantities of solids. Continuous screens have a guide rail at the channel bottom to support the screen elements. Construction of a recessed notch or step in the channel at the screen bottom is a good practice that can help prevent buildup of grit and debris ahead of the unit. Continuous screens are designed to be pivoted up and out of the channel for maintenance. Some screen systems include spray bars and brushes to improve cleaning of screen elements. A disadvantage of continuous screens is possible solids carryover resulting from the front-clean/back-return design. Another disadvantage is the removal of a high percentage of organic material, which makes storage and disposal of screenings more difficult, especially for unsupervised pumping stations.

1.1.4.5 Overhead Trash Rakes

Overhead trash rakes provide screening and conveyance equipment in one system. These systems are best suited for deep reaches beyond 15 m (50 ft) that may be of a depth difficult for traditional mechanical screens to be cost-effective. Components of this screen system include a monorail track system, cable-driven rake, and a stationary bar rack. The rake is lowered by an overhead hoist mounted to a trolley on a monorail system. The single rake is lowered into the channel where it engages the bar rack to remove debris. The rake is then pulled up by the hoist to a home position at which point the trolley is engaged to travel along the monorail. A typical system has disposal containers that are set under the monorail track so that the rake can then open above and drop screenings into the container. The monorail system is not limited in travel distance, but it is best practice to design shorter travel lengths to shorten the bar rack cleaning cycle and to reduce costs.

Considerations for operations include the inability to provide a continuous cover for the screening's container and the potential for screenings to fall out of the rake during the travel to the container.

1.1.5 Comminutors or Grinders

Comminuting devices are installed in the wastewater flow channel to grind and shred material into sizes from 5 to 20 mm (0.25 to 0.75 in.) without removing the particles from the flow. The best application of comminutors is in small outlying pumping stations on wastewater collection systems, where problems with large solids or rag balls or lack of options for screenings maintenance and disposal exist. The use of such a device is intended to reduce odors, vectors, and unsightliness often associated with screenings, prevent the buildup of screenings within the wet well, and reduce instances of pump failure due to clogging.

Although an older technology, comminutors have evolved to offer several configurations such as drum-type, straight-through, inline, dual-shaft, and shaftless.

Unfortunately, solids from comminutors have caused downstream problems at WRRFs. Pulverized synthetic materials can pass through some fine screens and will not decompose in the sludge digestion process and, if allowed to remain in the sludge, could bar public acceptance of the sludge for reuse as a soil amendment. Comminutors may create "ropes" or "balls" of material (particularly rags) that can clog mechanical aerators, mixers, pump impellers, sludge pipelines, and heat exchangers at WRRFs. Because operational experience with comminutors has been generally unsatisfactory, they are avoided in new designs and are being removed from many existing facilities.

For ultimate disposal in a sanitary landfill, grinding of screenings is typically unwarranted. Should there be any constraints (space, financial, structural, or electrical) in the pumping station that prevent the installation of a grinder(s), then pumps with chopping capability (chopper pumps) are recommended.

1.2 Screenings

Both the quantities and characteristics of screenings to be collected and removed require careful consideration before design of screening systems.

1.2.1 Quantities of Screenings

The quantity of coarse screenings to be removed can vary significantly, depending on the bar screen opening, flow characteristics of the municipality, type of collection system, and type of screen. These quantities are difficult to estimate without actual operating data. Data should be obtained whenever possible to design a successful operating system.

For separate wastewater systems, the most important variable affecting screenings quantity is typically the clear opening between bars. This feature should be adjusted by the design engineer according to the type of collection system and screenings characteristics.

For bar openings between 25 and 50 mm (1 and 2 in.), the volume of screenings removed per unit of flow is proportional to the clear opening dimension. In this range, for each 13-mm (0.5-in.) reduction of clear opening size, the volume of screenings will approximately double. For screen openings less than 25 mm (1 in.), the volume of screenings removed will increase more rapidly as the clear opening decreases to approximately 6 mm (0.25 in.).

The quantity of screenings removed will partly depend on the length and slope of the collection system and whether pump stations exist. Screenings quantities may be greater with a short, gently sloping collection system with low turbulence than with lengthy, steep interceptor systems or systems with pumping stations. This difference stems from disintegration of solids exposed to long-duration turbulence. The effect of turbulence will be more significant for smaller bar openings of less than 12 mm (0.5 in.), which capture more organic solids.

Climate can also play a role in the quantity of screenings captured. Long periods of dry weather can allow a buildup of materials within the collection system that get "washed" to the facility during storm events. This surge of screenings is commonly referred to as a *first flush*. Screening quantities during these events are difficult to estimate and can overwhelm screens, conveyance equipment, and maintenance staff. Stormwater and combined sewer systems produce several times the coarse screenings produced by separate wastewater systems as trash, leaves, and other litter enter the system. In addition, the quantity of screenings removed in stormwater and combined systems during wet periods will greatly exceed that during dry periods and that of typical screenings generated by wastewater systems.

1.2.2 Characteristics of Screenings

Composition and volume affect the disposal of screenings. Generally, coarse screenings can consist of branches, bottles, bricks, cans, rags (nondispersibles), paper, sticks, leaves, food particles, bones, plastics, bottle caps, and rocks. Typically, anything large enough to enter a manhole or storm inlet and traverse the collection system piping can end up at the screens. With smaller openings near 6 mm (0.25 in.), cigarette butts, fecal matter, hair, and other organic matter are included in wastewater screenings. Screenings, typically containing approximately 10% to 20% dry solids, will have a bulk density ranging from 640 to 1100 kg/m^3 (40 to 70 lb/cu ft).

1.2.3 Handling of Screenings

The method of cleaning manual or mechanical screens relates to the method of removing and transporting screenings to a disposal site or container. When designing a manually cleaned screen, relatively shallow screenings channels (<2 m [<6.6 ft]) are more suitable, with a drainage plate to allow water drainage of the screenings before shoveling. The method used to carry the screenings to a truck or other transport may range from a wheelbarrow to a bin carried by an overhead crane or monorail to a conveyor belt system. A belt conveyor, when used, is typically interlocked to start when the screening device starts or when initiated by an operator and runs for several minutes after the screening device stops or is initiated. The conveyor would typically be used to discharge noncompacted screenings to a truck-loadable bin.

Compaction of the screenings removed is an option for mechanical dewatering and reducing the volume of screenings. This is used primarily for medium to large pumping stations. Such devices, including washing presses or screw compactors, receive screenings directly from the screens and discharge the compacted screenings a short distance to portable bins or conveyors. Screw compactors are not designed to push the screenings long distances and should not be considered as a substitute for a stand-alone conveyance system. Compactors can reduce the water content of the screenings by up to 50%, with a volume reduction of 6.5 to 1. However, a source of water is required by the screw compactor for efficient screenings washing within the equipment. As a good practice, controls should sense jams, automatically reverse the mechanism, and activate an alarm when the motor is overloaded. The increased quantity of solids removed by the smaller openings of continuous screens are handled well by screw compactors.

To complete the handling of screenings, the material must be transported to a disposal site in accordance with local, state, and federal regulations. Any transportation or disposal alternative must be sensitive to the travel distance to disposal site, quantity of screenings to be disposed of, remaining water in the screenings, and environmental regulations. Screenings are typically disposed of at municipal landfills.

Screenings stored in containers at the pumping station should be covered or enclosed within containers to reduce vector attraction (rodents and insects) and odors. Regulatory agencies may provide further guidance on specific requirements. Containers larger than residential trash bins are typically rented from waste hauling companies and can range in size from 6 to 40 cu yd capacities. Considerations should be made for truck access related to loading, unloading, and staging of containers. Container covers are available in several styles from manufacturers or suppliers. The most cost-efficient cover would be a vinyl tarp or plastic lid. More complex cover systems can include odor control piping connections and rolling or hinged openings. The human effort required to install and maintain a cover system should be considered as part of design.

1.2.4 Design Practice

In selecting screening devices, the following should be considered. When agitated, rags that are comminuted tend to agglomerate into a rag ball that looks like an oily waste rag. This is not a problem at a remote pumping station where only pump protection is required. However, at a WRRF, the rag ball could disrupt downstream treatment processes.

In bar racks, the smaller the opening between bars, the greater the quantity of screenings will be removed. Therefore, bar spacing should be small enough to protect the pump but as large as is feasible to minimize the amount of screenings that must be removed. If the only criterion is to protect the pump, it is suggested that the distance between the bars be less than one-third the size of the maximum solid that the pump can discharge.

In a small pumping station, the construction cost of fixed bar screen or comminutor installation may represent a relatively large percentage of the total station cost. Access to the screens and screenings container should be included in the layout and may pose challenges for small pumping station sites.

1.2.5 Installation

Screening devices should be installed in an outdoor or positively ventilated area in accordance with safety requirements and guidelines. In addition, devices should be in a well-lit area that is accessible to operating personnel. Water under pressure should be available for cleaning. Labor-saving equipment for handling screenings is also important. The entire screening installation should be designed to reduce odor release to the outside atmosphere and vector attraction.

Both manually and mechanically cleaned screen installations and comminutor screening device applications should include overflow weir bypass channels or equivalent arrangements that function automatically to limit head differential on the screening device and prevent channel or upstream manhole flooding in case of screen-cleaning failure or excess hydraulic flow. Multiple screening devices are desirable in most installations to permit satisfactory screening over the full flow range to which the station is subjected. As flow to the station increases beyond the capacity of one screening device, a second device should be put in service—automatically, if possible. Provisions should be made for isolating screening units for maintenance.

The question of whether screening equipment should be located in an enclosed structure depends on two conditions: the equipment design and the climate. In climates with freezing temperatures, a heated enclosure is necessary. A structure will not only protect the equipment but will also ease maintenance and improve aesthetics. In climates with low-humidity and hot temperatures, screenings will dry quickly on bar rakes and conveyance mechanisms that may require periodic cleaning to reduce buildup. In windy areas, the screen rake area and the discharge chute area need covers to prevent screenings from blowing away.

1.2.6 Design

The first consideration in selecting a bar screen for a specific application is determining the maximum normal daily flow and maximum possible storm or peak flow. In connection with flow, the following factors govern the selection of bar screens.

The velocity of flow ahead of and through a screen affects its operation. The lower the velocity, the greater the amount of screenings that can be removed. However, the lower the velocity, the greater the amount of solids will be deposited in the screen channel. Therefore, a compromise is necessary between a velocity that permits deposition and one that will permit 100% removal of material of a certain size. Selection of an average flow velocity of 0.6 m/s (2 ft/sec) and a maximum flow velocity of 0.9 m/s (3 ft/sec) for wastewater flows is common practice. A minimum velocity as near 0.6 m/s (2 ft/sec) as possible should be maintained at low flows. Where stormwater is to be handled, approach velocities of approximately 0.9 m/s (3 ft/sec) are desirable to avoid grit deposition at the bottom of the screens. Screens might otherwise become inoperative when most needed during storms.

The velocity distribution in the approach channel is important for efficient screen operation. A straight channel ahead of the screen ensures good velocity distribution across the screen and maximum effectiveness of the device. Use of other than a straight approach channel has often resulted in large parts of the flow passing through one side of the screen. The channel hydraulic design should ensure that the flow is uniformly distributed.

The second consideration is determining the proper spacing of bars. Bar spacing should be based on the size of solids to be removed. Clear openings in the range of 6.35 to 150 mm (0.25 to 5 in.) are available. Clear openings for mechanically cleaned screens may be smaller than manually cleaned screens. Coarse screen bar openings selected range from 6.35 to 25 mm (0.25 to 1 in.); approximately 50% of the respondents favor openings of 20 to 25 mm (0.75 to 1 in.). A 20-mm (0.75-in.) bar spacing can be too large because twigs, ballpoint pens, and other plastics can pass sideways through the bars.

The third consideration in selecting a bar screen is the size of the bars, which directly affects the efficiency of the screen. Bar size is dependent on the width and depth of the screen channel and the condition of the flow. It is common practice to use 8 mm × 50 mm (0.3 in. × 2 in.) bars up to 2 m (6 ft) in length and 10 mm × 65 mm (0.4 in. × 2.5 in.) bars up to 3.5 m (12 ft) in length.

Other considerations are the size of the channels and the number of screens. Duplicate or multiple bar screens can be provided in a lower capacity range in a single station instead of one large bar screen. Duplicate installations are generally desirable to provide a degree of redundancy. If several bar screens are installed, the screen channels should be so arranged that any unit can be bypassed for service or repairs.

Trash racks are now seldom used ahead of coarse screens or grinders, except for combined systems with significant quantities of stormwater and at prisons with heavy screening quantities. Trash rack bar openings generally range from 40 to 75 mm (1.5 to 3 in.) (Water Pollution Control Federation [WPCF], 1989).

Additional design considerations for mechanical bar screens include discharge height to accommodate screen conveying equipment, angle of screen, construction materials for overall unit, drive unit, drive motor size and type of enclosure, spare parts, provision of redundant screen or bypass manual screen, odor control, conveyance system, disposal container size, and head loss through the unit.

1.2.7 Instrumentation

For mechanically cleaned bar screens, the following types of controls can be used alone or in combination: manual start and stop, automatic start and stop by timer control, high-level switch, and differential head-level-actuated starting switch on the cleaning mechanism. The differential head-level control system may be pneumatic pressure sensing, ultrasonic level sensing, or direct pressure sensing.

Mechanical bar screen installations at a minimum should include high upstream level and screen start–fail alarms.

A well-designed screening system will interlock auxiliary equipment such as spray wash, conveyors, or screening compactors with the operation of the screen. This will ensure coordination among the process operations.

1.2.8 Hydraulics

Based on an assumption of steady, two-dimensional flow, clean water head loss through a bar screen can be calculated by using the Bernoulli equation. This equation states that the depth of flow plus the velocity head before the screen equals the depth of flow plus the velocity head after the screen, plus friction losses. Actual head loss varies with the quantity and nature of the screenings allowed to accumulate between cleanings. Head loss created by a clean screen may be calculated based on the flow and the effective area of the screen openings. The effective area of the screen openings equals the sum of the vertical projections of the screen openings.

Head loss through a clean or partially clogged bar screen can be represented by the following equation:

$$H_L = K(V_1^2 - V_2^2)/2g \quad (8.1)$$

Where

H_L = head loss,
K = friction coefficient,
V_1 = velocity through bar screen,
V_2 = velocity upstream of bar screen, and
g = acceleration due to gravity.

Head loss through a clean bar screen is indicated by Kirschmer's equation below:

$$H_L = \beta(w/b)\ 1.33\ h \sin \phi \quad (8.2)$$

Where

H_L = head loss,
β = bar shape factor (Table 8.1),
w = maximum cross-sectional width of bars facing upstream,
b = minimum clear spacing of bars,
h = upstream velocity head, and
ϕ = angle of bar screen with horizontal.

Manufacturers' catalogs offer further details and guidelines in the sizing and selection of bar screens in addition to *Design of Water Resource Recovery Facilities* (WEF, 2023).

TABLE 8.1 Kirschmer's Values of β

Bar Type	β
Sharp-edged rectangular	2.42
Rectangular with semicircular upstream face	1.83
Circular	1.79
Rectangular with semicircular upstream and downstream faces	1.67

1.2.9 Grit Removal

Grit deposition and subsequent removal are potential problems for stormwater and wastewater pumping stations. Stormwater and wastewater grit materials have a settling velocity greater than that of organic solids and consist of discrete particles. Such materials include sand, cinders, rocks, coffee grounds, metal and plastic objects, and other relatively large-sized non-putrescible organic and inorganic substances.

It is standard design practice to not accumulate or remove grit at stormwater and wastewater pumping stations. Channels and wet wells are designed to provide slopes and velocities that will transport grit through the pumping station without accumulation. Where grit deposition is allowed or planned, the design should include provisions for ease of grit removal by the operators.

Where grit removal at a pumping station is desired, the various methods used in preliminary treatment at WRRFs are applicable and are described *in Design of Water Resource Recovery Facilities* (WEF, 2023).

2.0 ODOR CONTROL

2.1 Nature and Origins of Odor

Odors from septic or industrial wastes at a pumping station are a nuisance to operations personnel. They may also be a nuisance to nearby residents, and associated pollutants could pose health hazards to operations personnel. In the past, odors commonly associated with collection, pumping, and treatment systems have been somewhat accepted by the public. Today, off-gases, whether a health and safety issue or a nuisance condition, are typically not tolerated by the local community. Consequently, the location of the station and age of the wastewater are factors in the consideration of odor control.

Most odor-producing compounds found in domestic wastewater result from anaerobic biologic activity, which consumes organic material, sulfur, and nitrogen found in the wastewater. Organic odors are typically the result of biological activity that decomposes organic matter and forms a variety of malodorous gases, including indoles, skatoles, mercaptans, amines, and other gases (WPCF, 1990).

Sulfur is present in human excreta, and sulfates are found in most water supplies. Sufficient sulfur is typically available in domestic wastewater in the form of inorganic sulfates and sulfides or organic sulfides (e.g., mercaptans, thioethers, and disulfides) and is used as a food source by anaerobic bacterium, resulting in the production of odorous gases or vapors.

Hydrogen sulfide (H_2S) is the most commonly known malodorous gas emanating from WRRFs. It is highly soluble (2800 mg/L at 30 °C to 5650 mg/L at 5 °C) in normal domestic wastewater. In addition to its rotten egg odor, H_2S can cause highly corrosive conditions and is an extremely toxic substance in higher concentrations. Walls and crowns of sewers or closed tanks often have droplets of attached spray or condensation on the surface. Water (such as condensed water vapor in the form of dew or droplets) saturated with H_2S as a result of bacterial action forms sulfuric acid, which is corrosive to lead-based paint, concrete, metals, and other materials. The toxicity of H_2S is on the same order of magnitude as hydrogen cyanide (HCN); death may result in a few minutes when one is exposed to an excess H_2S concentration. For a fuller account of H_2S hazards, see *Safety and Health in Wastewater Systems* (WPCF, 1983).

At a pH of approximately 9, more than 99% of hydrogen sulfide dissolved in water is in the form of the non-odorous hydrosulfide ion (HS); while at a pH of 5, only 1% is in the non-odorous form. If the pH is above 8, there is insignificant production of H_2S gas. Figure 8.7 presents a graphic representation of the relationship between pH and H_2S production.

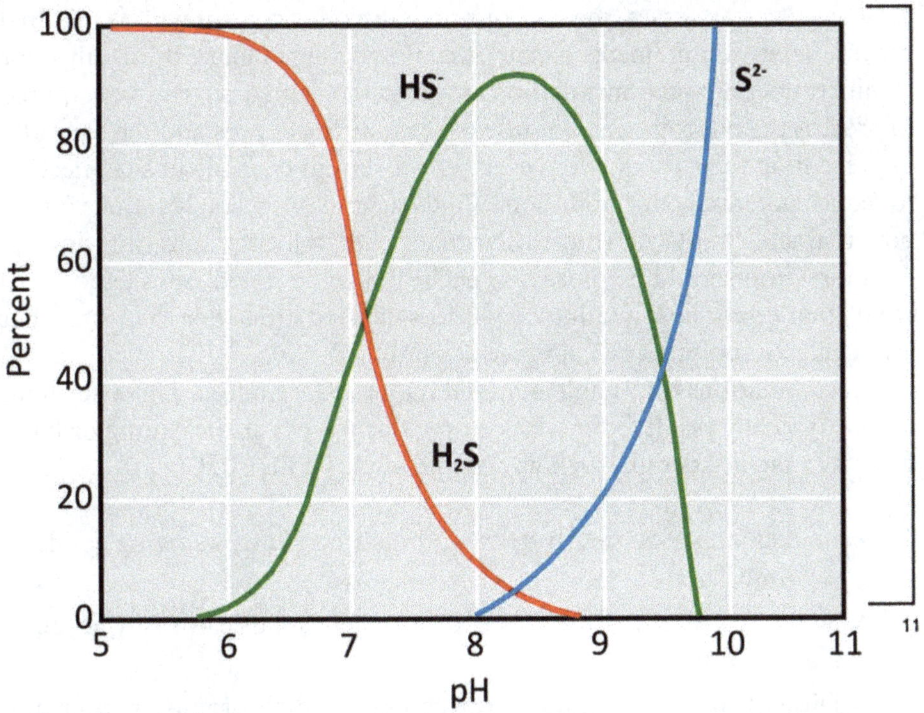

FIGURE 8.7 Effect of pH on Hydrogen Sulfide Equilibrium (Adapted from U.S. Environmental Protection Agency [U.S. EPA], 1985)

Hydrogen sulfide is produced by reduction of the sulfate ion (SO_4^{2-}) by anaerobic sulfate-reducing organisms. These bacteria thrive at low oxidation–reduction potentials (ORPs) of −0.20 to −0.30 V, pHs of approximately 6 to 9, and a temperature of approximately 30 °C. Anaerobic slime growths and sludge deposits that accumulate in wastewater lines typically have lower ORPs than wastewater, making them more conducive to H_2S production. Hydrogen sulfide may thus be produced even though wastewater contains up to about 0.3 mg/L of oxygen (O_2). The following section contains a brief description of odor control practices. For more information about odor control, refer to *Odor Control for Wastewater Facilities* (WEF, 2020).

2.2 ODOR CONTROL PRACTICES

2.2.1 Pumping Station Design and Operation

Designing for odor control requires that the engineer assess the effect of the odor on adjacent land use and workers at the facility. Significant criteria are frequency and intensity of occurrence related to the population affected. For example, the effect of an odor on a residential area will be substantially different from the effect on a heavy industrial area where some odor is already present. In certain cases, the cost of providing odor control will be minimal because safety requirements dictate effective venting of units; thus, only odor elimination equipment installations to treat the vented air will be required to process the odorous gas. Because of the nature of odors and the difficulty in preventing their production, every effort should be made in site selection to reduce potential odor pollution. Wind patterns at several locations on the potential site should be studied. Normal wind velocities and durations in each direction should be plotted at each spot to be tested. Sites exposed to winds that continuously dilute any odors emitted are better than sites that are regularly subjected to calm periods (WEF, 2020).

Accumulations of organic material may produce odors. These accumulations typically result from a lack of regularly scheduled cleaning or basic defects in facility design, such as the following (WEF, 2020):

- Inadequate velocities in open or closed conduits, allowing solids to separate;
- Square corners that provide pockets where solids can easily accumulate and are difficult to clean;
- Difficult-to-clean, rough surfaces upon which organic matter and slime can build up;
- Insufficient nearby hose bibs on pressurized water supply lines;

- Poorly designed drainage systems to carry flushing water cleanly away;
- Inaccessible locations that maintenance personnel cannot easily clean; and
- Inadequate lighting.

Stormwater pumping stations may also be sources of odors if poor maintenance and inadequate housekeeping permit the accumulation of debris, sludge, or trash or the buildup of organic slimes in the wet wells. Regular, careful cleaning is required of stormwater and wastewater pumping stations. The design engineer must provide facilities that permit and facilitate cleaning.

Adequate ventilation prevents obnoxious concentrations of odorants. Vented air should not be recirculated unless odorants have been removed. When untreated odorous air is recirculated even for a short time, odorants may be absorbed in the duct system. Odorous air must be treated or discharged in such a fashion that it will not cause a nuisance either at the facility or to neighbors of the facility.

Whenever odorants accumulate, there is a possibility that a hazardous gas is also present. This makes reliable ventilation even more important in hazard-susceptible areas. For safety, a flashing light or other similar warning device indicating a lack of air movement should be located at each entrance to a susceptible area; a sign telling what to do if the light is flashing should be posted nearby. This warning device should be triggered by an air movement indicator—not connected to the electric circuit of the fan to be triggered by the on–off electric current. Outside fresh air typically should be forced into the ventilated area, pushing the odorous air out through vents at either the top or the bottom of the room, depending on the specific gravity of the gas to be removed. Recently approved National Fire Protection Association (NFPA; 2020) 820, *Standard for Fire Protection in Wastewater Treatment and Collection Facilities*, requires fans on intake and suction. Hydrogen sulfide, for example, is heavier than air and is found near the floor, whereas ammonia (NH_3), which is lighter than air, would be found at the ceiling. In certain configurations, pump stations may be classified or have areas that are classified as Occupational Safety and Health Administration (OSHA) hazardous confined spaces. Workers and or other entrants to these areas should use proper hazardous gas detection equipment such as multi-gas monitors.

The design engineering should always investigate the concentration of H_2S at the beginning of the project because a substantial concentration of H_2S (more than approximately 1 mg/L) or a combination of high wastewater temperature >22 °C (>72 °F) and long residence time (>5 hours) in the wastewater may

- Be a deciding factor in choosing variable-speed drives;
- Require a sealed wet well or pump sump;
- Alter the location of the pumping station site;
- Require odor control units on the ventilation exhaust system to meet regulatory requirements;
- Increase the capital and operation costs of a ventilating system;
- Influence construction details, such as the kind of cement used or the type and lining of the wet well and sewers; and
- Require a specific regimen of washdown operations to keep acid-producing bacteria in check (Jones & Sanks, 2008).

Hydrogen sulfide investigation may be performed using deployable H_2S loggers, many of which can be left in a pumping station or wet well for multiple days. Considerations should be made to capture diurnal and seasonal changes in odor concentrations.

2.2.2 Odor Prevention

Because most odors generated in and around pumping stations are the result of anaerobic biological decomposition, eliminating or reducing these anaerobic conditions can be an effective way to prevent odors.

2.2.2.1 Flushing/Pigging

Regular flushing can be used to control organic deposits, sedimentation, slime growth, and resultant odor production. If the force main from the pumping station has a long residence time and is a source of odor production, a program of regular swabbing or "pigging" of the force main may be used to clean the line, thereby arresting the slime growth that leads to odor production. An added benefit of improved pumping efficiency may result from force main cleanings. Swabs or pigs tend to be bullet shaped of various materials and outer surfaces to match the specific conditions of the pipe material and type of cleaning to be performed. Pumping stations can be designed or retrofitted to facilitate pig launching. Figure 8.8 shows a typical pig-launching arrangement.

2.2.2.2 Air/Oxygen Injection

Improvement in the oxygen balance can be achieved by the addition of air or oxygen to wastewater. Addition of sufficient dissolved oxygen can prevent

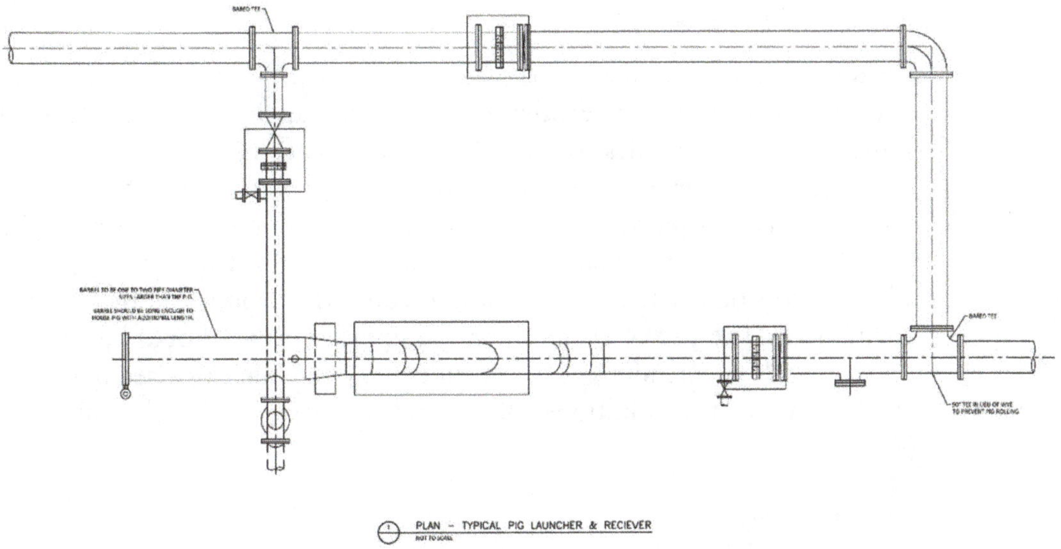

FIGURE 8.8 Typical Pig or Swab Launcher for Force Main Cleaning

or significantly reduce further sulfide generation from occurring and allow biochemical oxidation of dissolved sulfides. Methods of air and oxygen addition include the following:

- Aeration of wet wells;
- Direct injection of compressed air to force mains;
- Use of Venturi aspirators in force mains or lift stations;
- Use of air lift pumps at lift stations;
- U-tube dissolution using either compressed air or Venturi aspirators in either gravity lines or force mains;
- Pressure tank air dissolution systems for gravity lines (U.S. EPA, 1985); and
- Super oxygenation.

Super oxygenation systems provide a more concentrated dosing of pure oxygen into the force main or conveyance pipeline. These systems typically include a sidestream pump that pulls a portion of the raw sewage from the pipeline where the sidestream is injected with oxygen and mixed through a contactor, which optimizes the saturation of the oxygen in the sidestream. The oxygen-rich sidestream flow is then discharged back into the conveyance pipeline to increase the dissolved oxygen in the pipeline and reduce or eliminate anaerobic conditions.

2.2.2.3 Chemical Addition

Chemical addition under the right conditions can prevent the release of odorous substances. For example, alkalis such as lime can be added to water containing sulfides to raise the pH, thus keeping the sulfides in the ionized form rather than the H_2S form. Other chemicals are added to oxidize odorous molecules into more stable, odor-free molecules or raise the ORP and prevent the chemical reduction of sulfates. Chemicals are also added to kill or inactivate bacteria that cause the production of odorous compounds. The use of chemicals at pumping stations should be coordinated with downstream treatment facilities because their operation may be affected by chemical additions. Often-used chemicals include the following (WEF, 2020):

- Oxidizers
  - Ozone
  - Hydrogen peroxide
  - Chlorine
  - Sodium hypochlorite
  - Potassium permanganate
- Chemicals that raise ORP
  - Oxygen
  - Nitrate
  - Hydrogen peroxide
  - Chlorine
- Bactericides
  - Ozone
  - Chlorine
  - Hydrogen peroxide
  - Potassium permanganate
  - Chlorine dioxide
  - Sodium hypochlorite
- pH modifiers
  - Lime
  - Sodium hydroxide
- Precipitants
  - Trivalent metals such as iron and aluminum

2.2.2.3.1 Chlorination

Chlorination represents one of the oldest odor control practices. It is effective in odor control for two basic reasons: it is a strong oxidizing agent and it is an effective bactericide.

Chlorine reacts readily with many compounds in wastewater because it is a strong oxidizing agent. One of the most important of these reactions is with H_2S. Chlorine has been used successfully to destroy sulfides and prevent sulfide from building up in trunk sewers, pumping stations, and force mains. Hypochlorite feed systems are typically used when the requirement for chlorine is less than 2.3 kg/d (5.0 lb/d). If the dosage were 10 mg/L, this would correspond to chlorinating a flow of approximately 230 m^3/d (60 000 gpd). Equipment required for hypochlorite addition includes a hypochlorite storage vessel, metering pump, and diffuser or injector.

Chlorine is a hazardous material. Design of a chlorination station must recognize the need for the safe handling and storage of chlorine. Chlorine is also corrosive to equipment and ventilation requirements are greater. In some jurisdictions, an emergency chlorine gas scrubber system is required where chlorine gas is used.

2.2.2.3.2 Hydrogen Peroxide

Hydrogen peroxide (H_2O_2) is a weak disinfectant but a practical oxidant for odor control. Experience has shown that 1.5 to 4.0 mg/L of H_2O_2 is required to oxidize 1 mg/L of H_2S. For prevention of sulfide formation, dosage rates will vary with biochemical oxygen demand, temperature, and hydraulic characteristics of the sewer main or force main.

Some of the possible advantages or benefits of the use of hydrogen peroxide for sulfide control are as follows:

- It can be used for either gravity sewer or force main applications.
- The chemical feed system is relatively simple and inexpensive.
- Reactions with sulfide or other wastewater constituents produce non-toxic byproducts.
- The decomposition of excess H_2O_2 results in the addition of dissolved oxygen to the stream.
- With proper dosage, H_2S generation will be suppressed for 3 to 4 hours after H_2O_2 addition.

Adequate safety precautions must be observed in the handling of hydrogen peroxide. Protective clothing, including face shields, must be worn during bulk storage filling or repair or maintenance work where contact or spillage might occur. Any spills should be immediately washed down with water to prevent spontaneous combustion of organic materials in the presence of H_2O_2. Emergency eyewash and shower facilities are necessary for maintenance personnel.

Equipment required for hydrogen peroxide addition is relatively simple, consisting of a storage vessel, metering pumps, appropriate valving and transfer piping, and injection nozzle. For small or intermittent applications, a package dosing system can be used. In this system, hydrogen peroxide is withdrawn directly from containers delivered by the supplier.

2.2.2.3.3 Potassium Permanganate

Potassium permanganate ($KMnO_4$) is a strong oxidizing agent and may be used in aqueous solutions at various pH ranges to react with hydrogen sulfide and certain other odor-causing compounds. The actual reaction will depend on specific conditions. Therefore, dosages are case specific and difficult to predict.

Potassium permanganate has been used for sulfide control in collection systems and pumping stations, typically for small or intermittent applications. The relatively high unit chemical cost may be a limiting factor. Some advantages in using $KMnO_4$ are that it is easy to handle, relatively stable under normal storage conditions, and noncorrosive to most construction materials, and metering equipment is simple and easily installed.

As with hydrogen peroxide, adequate safety precautions must be observed in the handling of $KMnO_4$. Protective clothing, including face shields, must be worn during bulk storage filling or during repair or maintenance work where contact or spillage might occur. Emergency eyewash and shower facilities are necessary for maintenance personnel.

2.2.2.3.4 pH Adjustment

The addition of alkalis to wastewater can be used to adjust pH and effectively decrease the emission of H_2S gas to the atmosphere. The amount of H_2S present depends on the pH of the wastewater and the concentration of sulfides and precipitants.

Aside from the effect that increasing the pH has on the solubility of H_2S, alkalis also serve as good disinfecting agents at high concentrations, thereby inactivating the slime layer that forms along the interior of sewer walls. Sulfide generation in sewers can be stopped for several days if the slime layer is inactivated. If the pH is raised to 12, sulfide generation during the summer months may cease, but may return to 50% of the normal rate in approximately 1 week. If the pH is raised to 13, sulfide generation will practically cease for the first week, but it may return within 2 weeks.

The chemical dosage required to raise the pH to 12 makes odor control by continuous pH adjustment economically unacceptable; however, the infrequency of applying this dosage may prove to be competitive with other odor control methods in specific situations.

Systems operating successfully today primarily use caustic soda as the alkali. Lime, particularly hydrated lime, may also be used.

2.2.2.3.5 Metal Salts

The salts of many metals will react with dissolved sulfide to form metallic sulfide precipitates, thus preventing H_2S release to the atmosphere. For effective removal of dissolved sulfides, the metallic sulfide formed must be highly soluble.

Iron salts have been used for sulfide control because of economics and the low toxicity associated with iron. Water treatment sludges that are high in iron have been successfully used in controlling H_2S formation and, therefore, may be an attractive economic solution when available.

2.2.3 Odor Treatment

If prevention of odor generation is not feasible or cost-effective using the various techniques discussed previously, odorous air within a confined space can be removed and treated before being released to the surrounding atmosphere. Wet scrubbers, activated carbon, and other adsorptive or absorptive processes can be used to remove odor compounds from the air.

2.2.3.1 Wet Chemical Scrubbing

Wet scrubbing involves contact of odorous gas with a scrubber solution, typically in a countercurrent or crossflow fashion, to allow transfer of the odorants from the gas stream to the scrubber liquid by one or more of the following mechanisms:

- Condensation of odorous vapors,
- Removal of odorous particulates,
- Odor absorption to the scrubbing solution,
- Odor reaction with an oxidizing scrubbing solution, and
- Emulsification of odorous gases in a chemical reagent.

The most common scrubbing medium is water with chemical additives such as buffered potassium permanganate, sodium hypochlorite (NaOCl), caustic (NaOH), hydrogen peroxide, chlorine (CI_2), chlorine dioxide (CIO_2), or ozone (O_3). Potassium permanganate solution is often used as the scrubbing liquid. Sodium hypochlorite, on the other hand, has a faster reaction rate than $KMnO_4$ and is as effective. An important economic factor to take into account when considering $KMnO_4$ is whether the scrubbing liquid will be recirculated, thus obtaining maximum use of the oxidizing chemicals. Recirculation is typically cost-effective.

Scrubber water may have to undergo treatment before disposal. If operation is intermittent, accumulated sludge should be drained periodically. Scrubber water disposal is not a problem at a wastewater pumping station because the spent scrubber water can be introduced back to the wastewater stream at a rate that will not upset downstream processes.

2.2.3.2 Activated Carbon Adsorption

Activated carbon adsorption is a commonly used method for treatment of malodorous air. Adsorption is the phenomenon whereby molecules adhere to a surface with which they come in contact. Activated carbon has a high surface-to-volume ratio; thus, a large surface area is available for adsorption in a relatively small volume.

The quantity of material that can be adsorbed in a bed of activated carbon depends on the following factors:

- Concentration of material in the space around the activated carbon;
- Total surface area of the activated carbon;
- Total pore volume;
- Temperature;
- Presence of other competing contaminants;
- Characteristics of the compounds to be absorbed (molecular weight, boiling point, polarity, size, and shape);
- Polarity of the activated carbon;
- Relative humidity of vapor stream; and
- Contact time of the vapor stream within the activated carbon bed.

Activated carbon is typically regenerated thermally, although special activated carbon (normally specified for H_2S removal applications, impregnated with caustic [NaOH]) can be regenerated chemically using a 50% solution of NaOH. Thermal regeneration using multiple-hearth furnaces requires removal of the carbon from the adsorption system.

The design of activated carbon adsorbers involves four fundamental steps:

- Characterize the contaminated air (volumes, concentrations of constituents) and the desired effluent characteristics;

- Select the adsorbent;
- Conduct pilot studies to determine expected performance, useful life of carbon, and design criteria; and
- Apply pilot data to full-scale design.

The importance of full characterization of the contaminated air cannot be overemphasized. Of particular importance is the diurnal variation in air volumes and odor intensity. Failure to accurately estimate these parameters may result in poor performance or higher than anticipated carbon replacement or regeneration frequency. Characterization of specific constituents may also be important in design because special impregnated activated carbon now available may demonstrate improved performance over standard carbon.

2.2.3.3 Dry Chemical Scrubbers

Dry chemical scrubbers are similar in configuration to a carbon adsorption system in that they consist of a dry media within a fiberglass-reinforced plastic (FRP) vessel in which the odorous air and compounds are forced through to provide contact with the media. Odorous air is typically pulled through the vessel via a fan mounted to the top of the vessel. The media consist of a proprietary dry chemical media that are engineered to react with and oxidize H_2S and other odorous compounds and convert them to a water soluble and non-volatile polysulfide polymer. These systems typically include a layer of activated carbon media for polishing before the treated air is discharged through the top of the vessel. Similar to carbon media, the vessel should be sized based on odorous airflow and compound concentrations to provide sufficient media volume for contact time and to minimize frequent changeout of reacted media.

2.2.3.4 Ozone

Ozone is a strong oxidizing agent and disinfectant. Ozone is effective in treating atmospheric H_2S, but is not effective in controlling dissolved H_2S in solution. Because ozone is toxic to humans in concentrations as low as 1 mg/L, odorous gases should be drawn into a confined space for treatment. Ozone is an unstable gas and must be generated at the site. Because of the relatively high capital cost associated with ozone generation, consideration of its use is typically only applicable to large pumping stations.

2.2.3.5 Organic Media Biofilters

Organic media biofilter beds have been successfully used to remove odors from relatively small volumes of air containing odorous compounds. The simplicity and economics of organic media biofilters may make them particularly attractive for pumping station odor treatment. Contaminated gas is introduced to a bed of organic media typically consisting of wood chips, bark mulch, and compost through perforated pipe encased in stone located near the bottom of the bed. Alternatively, air distribution plenums can be used in lieu of stone and piping. The plenums discharge the odors evenly and allow for replacement of media without damage unlike the stone and perforated pipe, which typically need replacement after media removal. As the gas passes upward through the bed, the media substrate absorbs and oxidizes odorous gases such as H_2S, sulfur dioxide, and NH_3. Oxidation products are water, carbon dioxide, sulfate, and nitrate.

Equipment requirements for an organic media biofilter are minimal. A simple system consists of a perforated pipe network surrounded by gravel to prevent plugging of the perforations and is overlaid with suitable media such as sandy loam or compost. Air piping should be constructed of acid-resistant materials such as polyvinyl chloride (PVC) or high-density polyethylene. Valves may be used to direct odorous air to certain portions of the filter bed or bypass parts of the bed to allow for maintenance activities. For moist climates in which soil saturation may be a problem, an underdrain system may be used to remove excess water or the beds may be covered to prevent direct infiltration from rainfall. In dry climates, periodic watering and inlet air humidification may be required to maintain a suitable moisture content in the bed to sustain bacteriological life. This may be accomplished manually or by using an automated irrigation system and fine mist nozzles in the inlet piping.

The design of organic media filters depends on the nature and concentration of odorous compounds, airflow rates, target concentrations in exhaust gas, and characteristics of the organic media. Designs vary from simple filter beds with networks of perforated pipe to more elaborate systems with concrete berms, underdrain systems to collect excess water, and automated watering hardware.

Because of the many variables affecting design and performance, it is recommended that pilot studies be conducted before designing full-scale facilities (U.S. EPA, 1985).

2.2.3.6 Biotrickling Filters

Biotrickling filters have been successfully used in larger sizes at WRRFs for decades. In recent years, biotrickling filters have been scaled down for the comparatively lower airflows typically required at pumping stations and lift stations. Biotrickling filters consist of an odor fan, a biotrickling filter vessel, non-organic media, and an irrigation system.

The odor fan discharges odorous air into the bottom of the biotrickling filter where the air flows upward through the filter media and is exhausted from the top of the vessel. The vessel is typically made of FRP and is cylindrical in shape. The media are made of a polyethylene material of various sizes and shapes and are manufactured to maximize surface area and pore space for bacteriological growth. The media are resistant to degradation and typically last 3 to 4 times longer than organic media found in other odor treatment technologies.

The irrigation systems are used to shower the media to sustain bacteriological life and rinse away byproducts from the consumption of the odorous compounds. For the removal of H_2S, the irrigation system will use the recycled water from the sump of the vessel that typically has a pH between 1 and 2 promoting the growth of autotrophic bacteria. For the removal of reduced sulfur compounds, recycled sump water is not used for irrigation; single-pass fresh water should be used to promote growth of heterotrophic bacteria.

2.2.3.7 Odor Modification, Counteraction, and Masking

Odor modification is the name given to a process whereby two substances of given concentration are mixed, with the resultant odor of the mixture being less intense than that of the separate components. In some cases, odor would then be imperceptible.

Odor counteraction is the name given to the phenomenon whereby odor intensity is reduced by adding odor agents that are not chemically reactive to a malodor. Examples of pairs of odiferous gases that, in certain proportions, result in the mixture being odorless or nearly so are methyl mercaptan and eucalyptol, skatole and coumarin, and butyric acid and juniper oil.

Counteraction agents are available to counteract the following odor types:

- Phenolic,
- Aldehyde,
- Amine,
- Solvent,
- Mercaptan,
- Aromatic, and
- Organic, fatty acid.

Odor masking is the name given to a method by which the quality of a malodor is overcome by mixing it with a substance having a strong, pleasant odor. Thus, the pleasant odor effectively masks the malodor, resulting in a less objectionable odor.

Typical masking agents include heliotropin, vanillin, eugenols, benzol acetate, and phenylethyl alcohol. These are organic aromatic compounds that can be effective as short-term or emergency solutions for severe odor problems.

Odor modification, counteraction, and masking applications have been met with variable success for wastewater odors. They have often been applied as interim solutions or for short durations during periods of odor emissions. In some cases in which odor-masking agents have been used, resulting odors have been considered equally unpleasant. The problem is often further complicated by the subjective human response to odors. In general, prevention of odor emissions and positive control and removal of the odor-causing substance(s) are preferred.

3.0 NOISE CONTROL

3.1 Noise Characteristics and Sources

Careful consideration must be given to the noise problems in pumping stations. One of the biggest concerns in deciding the location for pumping stations is the effect of noise on residential neighborhoods. Interior noise levels should also be considered because of the safety of operations and maintenance personnel exposed to long periods of high noise levels.

The most common noise problems in pumping stations come from pumps, electric motors, fans, compressors and blowers, air-conditioning units, and engine generators. The design engineer should be able to obtain noise data from the manufacturer of any of these pieces of equipment.

The designer should request sound power level (PWL) octave band in dB from the manufacturer or sound pressure level (SPL) octave band in dB over a known distance. Sound pressure level can be calculated to dissipate over distance, and it can be A-weighted and logarithmically summed for SPL in dBA. All summations are logarithmic in acoustics. Background sound should be summed with pump station noise.

3.2 Applicable Codes

The design engineer should be familiar with the latest OSHA occupational noise exposure and local environmental control ordinances. The intent of these codes is to protect the worker and the surrounding neighborhood from the effects of high noise levels. (See OSHA and local noise ordinances for permissible noise exposures.)

If the permissible noise exposure is exceeded in the interior of any facility, hearing protectors shall be used by the worker to avoid permanent hearing loss.

Besides meeting state and local noise ordinances, the design engineer should consider additional noise control measures to prevent any adverse conditions in the community.

3.3 Noise Reduction Measures

Effective noise control is typically achieved by means of isolation, dissipation, or a combination of the two. This involves the use of silencers, acoustical enclosures, sound barriers, acoustical louvers, and acoustical linings. Additional noise attenuation is achieved by the type of building construction materials used. Manufacturer literature should be used to obtain sound attenuation for this type of equipment and sound transmission class ratings for construction materials.

During the design of the pumping station, it is important to know if sound attenuation will be used because additional space may be required to install this equipment, and attenuation systems may limit access for maintenance. It is possible that equipment must be completely enclosed in buildings, or that 3- or 4-sided walls are required around exterior equipment. The designer should avoid large wall openings (louvers, fans, roll-up doors, and windows) on walls close to property lines, or the designer should avoid wall openings and instead use roof openings (hoods, fans). Depending on elevation of adjacent properties, the roof openings may also need attenuation.

The designer should ask the attenuation manufacturer for sound data after the sound enclosure or silencer is applied to the equipment. If that is unavailable, the designer can directly subtract insertion loss and directly add back airflow generated noise. All these summations are performed with PWL and are direct summations rather than logarithmic for noise reduction calculations. If considered 1-, 2-, or 3-sided sound barrier walls, with reflective surfaces such as buildings or storage tanks, a noise calculation software should be used to calculate expected noise at the property line. If there is a binding local noise ordinance and large concerns for pumping station noise, it is prudent to record background noise for 24 hours.

4.0 BUILDING SERVICES

4.1 Heating, Ventilating and Cooling

Heating, ventilating, and air-conditioning (HVAC) systems are required in pumping stations to provide comfort and safety for operations and maintenance personnel, protect equipment, and help control odor emissions. This section of the manual provides the HVAC design engineer with some design criteria, design procedures, and equipment now in use for HVAC systems.

4.1.1 Codes and Standards

The design engineer should be familiar with applicable state, local, and federal codes such as mechanical, plumbing, fire and electrical codes, U.S. EPA and OSHA regulations, recommended standards for wastewater works ("Ten State Standards"), and the American Society of Heating, Refrigerating, and Air-Conditioning Engineers (ASHRAE) standards.

4.1.2 Cooling Systems

Design criteria for heating and cooling loads are based on guidelines found in ASHRAE standards, particularly the ASHRAE fundamentals on nonresidential load calculations chapter.

Heat gain in pumping stations typically is greater than the heat loss. Pump rooms are typically not air-conditioned. The ASHRAE fundamentals motor heat gain equation can be applied with a safety factor to the pump horsepower to size supply fans, typically sidewall propeller fans with intake hoods, to blow outside air on the motors. The heat will travel to the top of the room, in which louvers or hoods, commonly sized at 500 fpm, can be used to relieve the air. If the pump room is heated, energy codes starting in 2015 require the building to be insulated, intake openings require motorized dampers, and exhaust/relief openings require gravity backdraft dampers. When using mechanical ventilation for space temperature control, a 10 °F (−10 °C) temperature rise applied to the internal heat gain is appropriate; for example, for outside air temperature of 105 °F (41 °C), the temperature at the top of the room would be 115 °F (46 °C) with all firm pumps running. Supply fans that are low blowing on motors will help maintain the breathing zone temperature equal to outside air temperature. The air velocity will increase occupant thermal comfort.

If electrical gear is housed in a building, that space should be air-conditioned for temperature and humidity control. For motor control center and switchgear rooms, the load will be lower, and an owner may choose no redundancy or one-half or two-thirds redundancy. For electrical rooms with variable frequency drives, the room temperature will increase rapidly if the air-conditioning unit fails, so N + 1 redundancy is recommended. Ductwork may be required for adequate air distribution. Energy codes require systems over 4537 kg (5 tons) to have 100% outside air economizers or meet the air-conditioning equipment efficiency exception. The exception should be pursued for corrosive environments. A minimum amount of outside air is recommended to positively pressurize the room. For corrosive environments, the intake should be high on the roof, and the designer may choose to specify chemical filters, aluminum or FRP ductwork, coil coatings, or separate positive pressurization units.

4.1.3 Ventilation Systems

Ventilation systems in hazardous areas (wastewater wet wells) shall be designed for 100% positive outside air. These systems typically comprise fresh air intakes (louvers and gravity ventilators), ductwork, supply fans, exhaust fans (roof or inline), exhaust outlets, and controls.

Class I and Division 1 and 2 hazardous areas, as defined by NFPA, shall be monitored for combustible gases or toxic fumes such as methane, gasoline, and hydrogen sulfide. Ventilation rates shall be doubled on detection of 20% lower explosive limit for methane or gasoline or 10 mg/L for hydrogen sulfide. If the gas concentration increases, local and remote alarms shall be energized.

The designer may specify plastic fans and louvers (FPR, PVC, and polypropylene) and FRP ductwork for corrosive environments, coated aluminum or steel fans with motors located outside of the airstream, or aluminum ductwork.

4.1.4 Heating Systems

Heating systems typically consist of unit heaters, air handling units with supply fans, heating coils, filters, air dampers and exhaust fans, ductwork, and controls. A boiler (natural gas, oil fired, digester gas, or electric) typically uses an energy source. Self-contained gas-fired and electric units are also used as other sources of heat, particularly for small stations.

Unit heaters are sized to take care of heat transmission losses through the building envelope and infiltration load. Radiant heaters may be used for spot heating, or heat trace can be used.

4.1.5 Wet Wells

Proper positive ventilation design is important in wastewater wet wells because of the possible buildup of combustible and toxic gases (methane and hydrogen sulfide, respectively). Because of the possible presence of these gases, it is desirable to design the wet well so that it is completely isolated from the dry well. Design engineers should become familiar with the minimum ventilation rates required to control these gases to prevent fatal accidents.

The potential buildup of hazardous gases in wet wells of stormwater pumping stations is less likely than in the wet wells of wastewater pumping stations. Therefore, ventilation system criteria for wet wells in stormwater pumping stations are not as stringent.

All ventilation equipment in wastewater wet wells shall be classified as non-spark construction in accordance with the Air Movement and Control Association and all motors and wiring shall be classified as explosion-proof per the National Electrical Code (NEC).

According to the "Ten State Standards," the minimum ventilation for wastewater wet wells is 30 air changes per hour intermittently or 12 air changes per hour continuously. Also, refer to the chapter on ventilation in NFPA 820 (NFPA, 2020) and *Design of Municipal Wastewater Treatment Plants* (WEF, 2023) for recommended minimum ventilation rates.

4.1.6 Dry Wells

Heating and ventilating systems in dry wells of pumping stations are designed to remove excess heat from equipment in warm weather and maintain adequate space temperature for maintenance personnel in cold weather.

The heat gain in dry wells is produced by building envelope condition, which only applies for the above-grade walls and roof of the structure, pump motors, and variable-speed drives and other electrical equipment. The heat dissipation by the equipment often exceeds the structure heat loss and excess heat must be removed by a ventilation system.

The heat dissipated by the motors of the pumps will help in winter to decrease space heating requirements and conserve energy.

As per the "Ten State Standards," the minimum ventilation for dry wells of wastewater pumping stations is 30 air changes per hour intermittently or six air changes per hour continuously. Efficient ventilation is typically accomplished by forcing air into the bottom of the dry well and exhausting it through the roof. This system of ventilation displaces stagnant air from the dry well and promotes circulation to provide a better environment. Ventilation systems for dry and wet wells shall be entirely separated.

4.2 Electrical Design

Adequate illumination should be provided for all stations, with conveniently located light switches. Shadows and glare should be avoided in the vicinity of machinery and floor openings. Deep wells should have multilevel light fixtures. All fixtures should be located so that maintenance activities will not be impaired. It is necessary to have adequate lighting so that maintenance work may be performed in pumping stations. Because small power tools frequently are used for maintenance, electrical receptacles or a portable compressor for air-operated tools should be placed near mechanical equipment.

The following lighting levels are recommended: 540 to 1080 lx (50- to 100-ft candles [ft–cd]) for offices, 320 to 800 lx (30 to 75 ft–cd) for control areas, 160 to 540 lx (15 to 50 ft–cd) for equipment areas, and 11 to 55 lx (1 to 5 ft–cd) for outdoor areas. Many states now have lighting limitations based on watts per square foot. This limitation should be investigated before the start of the design.

Manufacturers' catalogs provide data on wattage, voltage, lumens, and coefficient of utilization. Using these data, room dimensions; wall, ceiling, and floor reflective percentages; and the number of lighting fixtures can be calculated for each area.

Each room or area should have a National Electrical Manufacturers Association classification rating associated with it. This rating will determine the type of lighting fixture to install. The manufacturer will indicate in the catalog what type of environment the fixture is suited for. Wiring associated with the installation of each fixture shall be in accordance with the NEC and all applicable state and local codes.

Each fixture should have a high power factor and high-efficiency ballast to reduce electrical energy costs. Automatic lighting control should be considered in areas that are periodically occupied. Night lighting should also be considered for areas that are not operator attended at night but are periodically checked by guards or other personnel.

4.3 Plumbing Design

4.3.1 Water Supply

Water is frequently needed in pumping stations for sink and toilet use, cleaning of the building, and maintaining landscaping. Wet wells need to be flushed and cleaned frequently to minimize grease buildup on walls. If wet well walls are not cleaned regularly, grease can break loose, be drawn into pumps, and clog impellers and check valves.

A potable water supply is desirable for all pumping stations. A positive separation (reduced pressure principle backflow preventor device) should be maintained between the potable system and any lines or equipment subject to contamination. Warning signs should be posted at all water taps not directly connected to the potable supply. Hose bibbs are desirable at pump pits and screen rooms. If planting around the station exterior requires water, a freeze-proof hydrant for connecting a hose will be useful. Many pumps require seal water to flush the pump shaft and cool the packing within a stuffing box.

Clean water should be provided for sealing stuffing boxes of volute-type pumps wherever possible. Because a connection between the pump stuffing boxes and a potable water supply is a possible source of contamination, some means of preventing backflow of wastewater to the potable water supply is required. Wastewater may be contained if there is a positive break with an air gap between the potable water supply and the water seal system. Backflow devices consisting of a system of check and relief valves that protect against backflow may also be used. This latter method, however, is not approved by some regulatory agencies.

In water seal systems with an air gap, potable water is typically discharged to a constant-level tank that is located above any possible flood level. Some public agencies require that the potable water line terminate not less than 150 mm (6 in.) above the top of the tank. Discharge of potable water to the constant-level tank typically is controlled by a float valve. The required gland seal pressure may be maintained by locating the tank high enough in the structure to create the required head by gravity or by repumping from the tank.

If a pumping system is used, the water seal pumps may discharge either directly to the water seal piping or to a pneumatic pressure tank. In direct pumping systems, a pressure-relief valve customarily controls maximum system pressure and returns excess water to the sump tank. Direct pumping systems either are designed to run continuously or are electrically interlocked to run when any wastewater pump operates. With continuous pumping, energy use is high. This system is typically not recommended. For systems discharging to a pneumatic pressure tank, seal-pump operation typically is controlled from the tank pressure. Pump operation will be intermittent and recirculation losses will be eliminated with this system, which also reduces power consumption and possibly reduces water use.

Where conserving water is desirable or where graphite-impregnated pump packing is used, solenoid valves interlocked with pump starters may be installed in individual water seal lines so that water is supplied to a unit only when the pump is running. However, pumps having a flooded suction with the seal subject to internal pressure when the pump is not operating should be provided with continuous seal water supply.

A supply of approximately 0.03 to 0.06 L/s (0.5 to 1 gpm) for small mechanical and packing seals and 0.12 to 0.2 L/s (2 to 3 gpm) for large mechanical and packing seals has been found to be adequate for estimating water seal supply requirements. A generally accepted practice has been to provide water seal pressure at seals not less than 34 to 103 kPa (5 to 15 psi) greater than the pump shutoff head. Consult pump manufacturers for recommended water seal supply and pressure. Some pumps have a seal water system that is self-contained. They discharge water and pipe it back to the stuffing box.

The maximum water pressure for most buildings, associated plumbing fixtures, and devices is 80 psig per code. If the water pressure exceeds 80 psig, then a pressure regulator device will be necessary on the building water service.

A portable water supply is required for all sinks, restroom lavatories, and, if provided, also for all emergency showers and eyewash fixtures. Potable water is preferred to all hose bibbs and freeze-proof wall hydrants, but not required if appropriate non-potable water signage is provided at each of these devices receiving non-potable water.

Water heating systems are required for handwashing sinks and lavatories per most if not all codes. Typically, codes will require water temperatures for handwashing to be 105 to 110 °F (41 to 43 °C). For emergency use fixtures, if water temperatures cannot be maintained at between 60 and 90 °F (16 and 32 °C) without a water heating source, then one will be necessary. Some areas may have an incoming water temperature in that temperature range and may not require water heating for emergency use fixtures.

4.3.2 Interior Sanitary Drainage

Drainage sumps should be provided in dry wells for collection of water for pump packing or seal leakage, condensation or dehumidification, waste, piping leakage, groundwater leakage, and wastewater leakage—either inadvertent or from dismantling of pumps or piping.

Pump stuffing boxes are typically connected to drain by 13-mm (0.5-in.) pipes. Air-relief valves or air discharges from self-priming pumps should have piping appropriately sized to a drainage gutter or sump.

To ensure adequate building drainage, a minimum floor slope of 1 cm/m (0.12 in./ft) should be provided. The station also should have sufficient floor drains in both size and quantity. These drains should be located at the lowest spots or in places where spent water may easily be collected.

Dry-pit drainage sumps should be equipped with automatic level-actuated duplex sump pumps, when a gravity drainage system is not possible. Discharge piping is typically directed to the wet well and must extend above the maximum flood level, with provisions for prevention of backflow by siphon-breaker or double check valves to prevent accidental flooding of the dry pit.

Drainage from restrooms, shop sinks, mop sinks, general floor drains, and so on should be provided when applicable. These drainage systems should be designed to drain by gravity (no pumps necessary) when possible.

When a gravity drainage system is not possible because depths of mains are an issue, a lift station may be necessary.

Sanitary waste drainage may require a septic system if city or county sanitary sewer service is not on-site or nearby.

4.4 Site Storm Drainage

Pump station site should be graded to drain. A 3-to-1 maximum slope is preferred to allow adequate maintenance of the location around the pump station. However, a minimum 2% drainage should be provided for the area surrounding the pumping station building to prevent ponding at the pumping station site.

4.5 Utilities

It is necessary to contact each utility early in the design to obtain individuals' requirements and the name of the representative assigned. All telephone conversations should be backed up with confirmation emails or some other form of documentation. A site visit with each utility representative is recommended.

4.5.1 Electric Utility

Provide the utility with the following information: a load summary of the pumping station, both connected and normally operating; an aerial photographic survey or site plan to mark the location of the pumping station; and the preferred point of service. Discuss with the representative the starting methods and frequency of starts for the proposed pumping station and the location for the point of service.

Obtain the following information from the utility company: a complete current rate schedule applicable to the pumping station, the available short circuit current from the utility power line, transformer specifications, size, impedance, insulation type (dry or oil), primary fuse size, and their policy on underground service. After analyzing rate structures, the demand charges, power factor clauses, and underground service installation, discuss with the representative the advantages and disadvantages of each. Discuss transformer ownership and the utility's maintenance policy if the utility is providing the transformer. All underground service cables must be encased in red-dyed concrete. Determine whether the costs for the underground service will be shared by the utility company or if the owner is responsible.

Provide the following construction information to the electric company: specifications for any special switchboards, substations, and grounding requirements; site plans; building power plans; single-line diagrams;

substation plans; and elevations for service entrance equipment. Also, provide the approximate date of service and the bid opening date.

4.5.2 Telephone Utility

Obtain an agreement with the telephone utility on the following: approximate date of service; service point; routing of telephone company cables; the location of its equipment in the owner's facility; a clear definition of items and interface furnished by the telephone utility and the owner; telephone terminal facilities (e.g., plywood panel and enclosure); telephone extension conduit size and type; telephone extension outlet box size, type, location, and mounting height; and equipment interface requirements (e.g., telemetry and surge suppressors).

Provide the following construction information to the telephone utility: specifications for the telephone terminal cabinet, site and building plans containing the telephone requirements, the approximate date of service, and the bid opening date.

4.5.3 Gas Company

The following information should be obtained from the gas company: gas meter location, gas pressures, maximum cubic feet per hour of gas supplied, and size of meter. Provide the utility with the location of gas piping for connection to the main gas pipe, along with estimated design loads.

5.0 OTHER APPURTENANCES

5.1 Hoists

Hoists, either manually or electrically operated (explosion-proof in areas classified Class I, Division 1 and 2) and either fixed or portable, should be provided in all stations for the handling of equipment and materials that cannot be lifted readily or removed from the station by manual labor. Moreover, it has been found that hoisting facilities typically pay for themselves in the handling of equipment and materials in the original installation in large stations. It may be advisable to provide overhead bridge cranes and electrically operated hoists in large stations or where frequent use is anticipated. Hoists should be provided over basket screens, submersible pumps, other pumps, or motors, and in other locations where it is necessary to lift pieces of equipment or containers that cannot be handled by two people. Providing heavy lifting "eyes" in the ceiling over pumps and heavy valves and equipment will substantially ease the close handling of such equipment.

To avoid accidents and injuries common to the use of ladders or rungs, personnel elevators are desirable in deep stations. Elevators are typically regulated or certified by local, state, or federal ordinances or laws. Elevators in unattended pumping stations should be equipped with an alarm system connected to a location that is always attended.

Every piece of equipment installed in a pumping station can be expected to be removed during the life of the station. Hoist/crane capacity should consider the wet weight of all equipment with a 1.25 or 1.5 safety factor. All equipment to be removed should be equipped with lifting lugs to allow safe pull and lift of equipment. When equipment, such as pumps or motors, must be raised to the ground floor or the loading dock floor, removable grating placed immediately above the pump on each floor through which equipment must travel greatly eases equipment removal. Overhead beams centered above equipment to be removed and equipped with chain hoists or electric hoists are equally important in facilitating equipment removal. Once equipment has been raised to the level of the ground floor or loading dock floor, it is useful to have a beam or monorail leading through double doors to where the equipment can be loaded onto a truck from a loading dock.

Overhead hooks, tripods, and A-frames are useful in disassembling equipment, but they are inadequate for equipment removal. Pivoting cranes and monorails are convenient accessories for the installation and removal of equipment.

As previously mentioned, it is important to provide sufficient room within the station for equipment repairs to be performed and parts to be removed.

Planning also must include easy means of removing installation equipment from the pumping station. For small stations, A-frames and dollies are typically sufficient. For large stations, however, a monorail or a traveling crane may be needed to remove heavy items.

5.2 Safety Features

Accidental spills of toxic or flammable liquids—such as paint thinner from an industry or gasoline spills or gasoline tank overflows that enter sanitary or combined sewers—may lead to the release of explosive vapors in sewers and pumping station wet wells. Natural gas main leaks often cause explosive gas to enter sewers through pipe cracks or joints. Gas may follow the flow to wet wells and cause explosion hazards. Because dry wells are supposed to be separated from wet wells, the danger that explosive vapors may enter a dry well is reduced; nevertheless, potential danger still exists because of possible leakage around pumps.

Mechanical ventilating systems with ducts extended to within 0.6 m (2 ft) of the bottom floor level in dry wells are effective in minimizing the danger of explosions in these confined spaces. (Chapter 6 presents a discussion of explosion-proof motors and controls.)

The manual, *Safety and Health in Wastewater Systems,* discusses confined-space entry and safe working procedures (WPCF, 1983). Safety equipment should be specified from the beginning by the design engineer so that it is included as part of the project's construction costs. If it is assumed that this equipment will be bought later from operating revenues, the safety program should be initiated immediately at the beginning of the operation.

As noted earlier, considering the frequent necessity to enter pumping stations, it is in the interest of safety for the design engineer to use stairs in place of vertical ladders. Falls from ladders and other devices are the second largest cause of compensation injuries. Ladders that are not permanently anchored are inherently hazardous. However, permanently anchored ladders in wet wells are susceptible to corrosion and should not be used.

Even though confined-space safety procedures require standby workers to be present outside a confined space when entry workers are inside, telephones and radio systems should be installed permanently in buildings or trucks so that help may be used in an emergency. Pumping station emergencies typically require prompt attention.

The following safety features should be included in all stations:

- Railings should be provided around all manholes and openings where covers may be left off during operation and at other locations where there are differences in levels or where there is danger that the operator may fall.
- Guards should be placed on and around all mechanical equipment where the operator might come in contact with (or clothing might become entangled with) belt drives, gears, chain drives, rotating shafting, or other moving parts of equipment.
- Warning signs should be placed near dangerous machinery, hidden obstacles, and other hazardous locations. Good safety practice rules should be posted in appropriate locations.
- Rubber mats of a suitable type should be provided in front of all electrical equipment where there is any hazard from electric shock. Electrical equipment and wiring should be insulated and grounded properly. Switches and controls should be of the non-sparking type. Adequate lighting should be provided in all locations, especially where there is moving equipment or openings in the floors. Wiring and devices in hazardous areas should be explosion-proof. All receptacles should be of the three-wire grounding type outlet, and a ground test of each should be conducted after installation. Portable power tools, extension cords, and trouble lights (including their outlets) should be

of the three-wire grounding type. Lighting toggle switches should be "T" rated. When sump pumps are used with a flexible plug-in power cord, the cord and attachment cap should be of the three-wire grounding type. The attachment cap should be locked to the outlet with a locking device and the outlet should be protected by a combination toggle switch with built-in overload protection (fractional horsepower manual starter). It is good practice to provide a prominent sign on the major switchgear stating the service voltage. Disconnect switches for equipment located out of sight of the main panels, or motor control centers should be provided such that the operator may lock out energizing equipment when performing maintenance.

- Stairways should be used instead of ladders or manhole steps, with straight-run stairways preferable to the circular ones. Stair treads should be of the nonslip type. Vertical ladders should be provided with safety cages. Vertical ladders higher than 2.4 m (8 ft) should be provided with intermediate platforms or landings. Service elevators are desirable in larger stations with relatively deep pump pits. The use of manhole steps should be restricted to areas of infrequent use and, if at all possible, to level changes not exceeding 2.4 m (8 ft). Grab bars should be provided approximately 1.2 m (4 ft) above the level of the landing adjacent to the top step.
- The possibility of slipping on oil or biological growths that may develop near leaks presents a considerable safety hazard. Thus, proper drainage is important.
- A telephone is desirable because it will permit the operator to maintain regular contact with the main office. Also, in case of injury, fire, or equipment difficulty, it will enable the operator to obtain assistance as rapidly as possible.
- Testing and monitoring equipment should be available for checking the gaseous content of the atmosphere (particularly in the wet well), oxygen levels, and toxins. Explosive mixtures may also be encountered because it is almost impossible to eliminate completely the discharge of flammable substances to sewers in large systems. Mechanical ventilating systems in both wet and dry wells will reduce the danger of explosions in confined spaces.
- Suitable fire extinguishers should be located strategically throughout the station. The use of carbon tetrachloride types should be avoided because of the toxicity of this liquid and its vapor.
- Safety harnesses with ropes, first-aid supplies, lifesaver floating rings, hard hats, and oxygen respirators should be either stored at the

pumping station or furnished to each work crew. In the design and construction of pumping stations, all safety requirements must follow prescribed local, state, and national safety laws and codes.

6.0 REFERENCES

Jones, G. M., & Sanks, R. L. (2008). *Pumping station design* (3rd ed.). Butterworth-Heinemann.

National Fire Protection Association. (2020). *National Fire Protection Association 820: Standard for Fire Protection in Wastewater Treatment and Collection Facilities.*

U.S. Environmental Protection Association. (1985). *Design manual odor and corrosion control in sanitary sewerage systems and treatment plants* (EPA-625/1-85/018). U.S. Environmental Protection Agency, Office of Research Development.

Water Environment Federation. (2020). *Odor control for wastewater facilities* (WEF Manual of Practice No. 25).

Water Environment Federation. (2023). *Standards for design of water resource recovery facilities* (6th ed.; WEF Manual of Practice No. 8).

Water Pollution Control Federation. (1983). *Safety and health in wastewater systems* (Manual of Practice No. 1).

Water Pollution Control Federation. (1989). *Design practice survey.*

Water Pollution Control Federation. (1990). *Operation of municipal wastewater treatment plants* (WEF Manual of Practice No. 11).

7.0 SUGGESTED READINGS

Pankratz, T. (2019). *Screening equipment handbook for industrial and municipal water and wastewater treatment.* Technomic Publishing.

Water Pollution Control Federation. (1989). *Wastewater treatment plant data survey.*

Index

CPSIA information can be obtained
at www.ICGtesting.com
Printed in the USA
LVHW050803090223
738835LV00001B/1

9 781572 78436